SpringerBriefs in Environmental Science

SpringerBriefs in Environmental Science present concise summaries of cutting-edge research and practical applications across a wide spectrum of environmental fields, with fast turnaround time to publication. Featuring compact volumes of 50 to 125 pages, the series covers a range of content from professional to academic. Monographs of new material are considered for the SpringerBriefs in Environmental Science series.

Typical topics might include: a timely report of state-of-the-art analytical techniques, a bridge between new research results, as published in journal articles and a contextual literature review, a snapshot of a hot or emerging topic, an in-depth case study or technical example, a presentation of core concepts that students must understand in order to make independent contributions, best practices or protocols to be followed, a series of short case studies/debates highlighting a specific angle.

SpringerBriefs in Environmental Science allow authors to present their ideas and readers to absorb them with minimal time investment. Both solicited and unsolicited manuscripts are considered for publication.

More information about this series at http://www.springer.com/series/8868

Vladimir Pacheco Cueva

An Assessment of Mine Legacies and How to Prevent Them

A Case Study from Latin America

 Springer

Vladimir Pacheco Cueva
Department of Culture and Society
Aarhus University
Aarhus
Denmark

ISSN 2191-5547 . ISSN 2191-5555 (electronic)
SpringerBriefs in Environmental Science
ISBN 978-3-319-53975-1 ISBN 978-3-319-53976-8 (eBook)
DOI 10.1007/978-3-319-53976-8

Library of Congress Control Number: 2017932111

Printed on acid-free paper

This Springer imprint is published by Springer Nature
The registered company is Springer International Publishing AG
The registered company address is: Gewerbestrasse 11, 6330 Cham, Switzerland

Acknowledgements

All of the efforts carried out to complete the fieldwork that led to the writing of this book deserve to be acknowledged.

I would therefore like to thank two of San Sebastian's community leaders, Jose Vicente Hernandez and Gustavo Mirio Blanco, for facilitating contacts with community members, artisanal miners, and water vendors.

I would also like to express my gratitude to those who assisted during fieldwork: Kelly Lamb, Greg Lamb, Elizabeta Busygina, Rafael C. Cartagena, Dr. Rune Dietz, Marvin Garcia (CARITAS, San Miguel); Morena Guadalupe Guzman, Rafael Pacheco Guevara, Gloria Ester (Tey) Hernandez Herrera, Vilma Joya, Gabriel Labrador, Dr. Barry Noller, Ramon Pacheco Pashaca, Dr. Jose Manuel Pacheco Paz, Saul Pacheco, Dr. Luis Alonso Reyes (Health Unit, Santa Rosa de Lima); Saul Antonio Rivas, Tania Sosa (Humboldt Center); Sigrun Müller Pallesen Schaumburg, Lara Schwarz, and Juan Ismael Ventura (Santa Rosa de Lima Parish).

Parts of the research carried out for this book were also used for the preparation of a report by the Salvadorean Human Rights Ombudsman (Procuraduría para la Defensa de los Derecho Humanos). I would like to express my gratitude to the team in charge of research and production of that report: Julio Quiñonez Basagoitia, Dr. Tara Van Ho, Edwar Josué Lizama Argueta, Pedro Cabezas, Yanira Cortez Estevez, and Xenia Marroquin.

The research for this book was possible thanks to the financial assistance of Cordaid, ASPRODE, SalvAide, and Aarhus University.

Contents

1   Methodology   1
    1.1   Research Technique   1
    1.2   Data Collection   2
    References   3

2   Knowledge About Mine Legacies, International Best Practice
    Standards and Mine Closure Regulation in the USA and El
    Salvador   5
    2.1   What We Know About Mine Legacies   5
    2.2   International Best Practice on Mining Closure and Reduction
        of Mine Legacies   7
    2.3   Mine Closure Regulation in the USA   8
    2.4   Mine Closure Regulation in El Salvador   9
    References   11

3   Social and Historical Remarks About Mining in San Sebastian
    and Assessment of the Mine's Current Legal Status   13
    3.1   Growth and Decline of the Mine   14
    3.2   Current Status of the Mine Owner   16
    References   19

4   Legacies of the San Sebastian Mine   21
    4.1   AMD as Primary Environmental Legacy in San Sebastian   21
    4.2   Socio-economic Legacies of the San Sebastian Mine   23
        4.2.1   Estimating the Market Price of Water in Relation
             to Household Income in San Sebastian   24
        4.2.2   The Cost of Obtaining Non-polluted Water in San
             Sebastian   26
        4.2.3   Health Problems in San Sebastian   28

4.3   Lost Opportunities   31
       4.3.1   Economic Opportunities Associated with Mining
                and Other Natural Resource Extraction Activities
                and the Impact of Abandonment   31
       4.3.2   Uncertainty in Relation to Land Tenure   35
4.4   The Growth of Artisanal and Small Scale Gold Mining
       (ASGM) in San Sebastian.   36
       4.4.1   ASGM and Its Status in El Salvador and Beyond.   37
       4.4.2   The Use of Mercury in San Sebastian.   38
       4.4.3   Mercury and Its Problems   40
References.   41

5   Reform, Awareness, Prevention and Remediation Strategies   45
    5.1   Legislative Reforms Needed to Conform to International
           Best Practice in Mine Closure   45
    5.2   Prevention and Awareness   47
    5.3   Remediating Legacies in San Sebatian   49
           5.3.1   AMD Remediation   49
           5.3.2   Improving Access to Non-polluted Water   51
           5.3.3   Finding Alternatives to Artisanal and Small
                    Scale Mining.   51
    References.   53

Conclusion   55

Appendix A   57

Appendix B   59

Bibliography   61

Abbreviations

ACMER	Australian Center for Mining Environmental Research
ADES	Asociación de Desarrollo Económico y Social de Santa Marta
AMD	Acid Mine Drainage
AMWG	Abandoned Mines Working Group
ANDA	Administración Nacional de Acueductos y Alcantarillados (National Administration of Aqueducts and Sewers)
ASGMI	Asociación de Servicios de Geológica y Minería Iberoamericanos (Association of Iberoamerican Geological and Mining Services)
ASPRODE	Asesoría a Programas y Proyectos de Desarrollo
AusIMM	Australasian Institute of Mining and Metallurgy
CAMMA	Conferencia Anual de Ministros de Minas de las Américas (Annual Conference of Mining Ministries in the Americas)
CEICOM	Centro de Investigación sobre Inversión y Comercio
CGC	Commerce Group Corporation
CMLR	Center for Mined Land Rehabilitation, University of Queensland, Australia
CNR	Centro Nacional de Registros (National Registry Office)
COCHILCO	Comisión Chilena del Cobre (Chilean Copper Commission)
DIGESTYC	Dirección General de Estadísticas y Censos (National Statistics and Census Directorate)
EHPM	Encuesta de Hogares de Propósitos Múltiples (National Household Survey based on the World Bank's Living Standards Measurement Study)
EPA	Environmental Protection Agency (US)
FINATA	Financiera Nacional de Tierras Agrícolas (National Fund for Agricultural Land)
Hg	Mercury
ICMM	International Council on Mining and Metals
ICSID	International Centre for Settlement of Investment Disputes
IFC	International Finance Corporation

ISTA	Instituto Salvadoreño de Transformación Agraria (Salvadorean Institute for Agrarian Reform)
IUCN	International Union for the Conservation of Nature
IVA	Impuesto sobre Valor Agregado (Value-Added Tax)
MARN	Ministerio de Medio Ambiente y Recursos Naturales (Ministry of the Environment and Natural Resources)
MCA	Minerals Council of Australia
MCMPR	Ministerial Council on Mineral and Petroleum Resources
MINEC	Ministerio de Economía (Ministry of the Economy)
MINSAL	Ministerio de Salud (Ministry of Health)
MMSD	Mining, Minerals, and Sustainable Development
MPRDA	Minerals and Petroleum Resources Development Act
NOAMI	National Orphan/Abandoned Mines Initiative
OCMAL	Observatorio Latinoamericano de Conflictos Mineros (Latin American Observatory of Mining Conflicts)
OIR	Oficina de Información y Respuesta (Freedom of Information Bureau)
RRA	Rural Rapid Appraisal
SCER	Standing Committee on Energy and Resources
SEC	Securities Exchange Commission
SMI	Sustainable Minerals Institute
SSGM	San Sebastián Gold Mining
UNEP	United Nations Environment Programme
USGS	United States Geological Survey
WB	World Bank
WHO	World Health Organization

Abstract

The study seeks to enrich the growing literature on mine legacies by examining a case study of a small abandoned mine in Latin America. Using a combination of Rapid Rural Appraisal and secondary source analysis, this study assessed some of the most damaging legacies of the San Sebastian Gold Mine in eastern El Salvador, compared the country's mine closure legislation against world's best practice standards and provided strategies for awareness, prevention, and remediation.

The most damaging legacy to the environment is that of acid mine drainage (AMD) contamination of the local river. The impact of AMD is felt well beyond the mining district and the costs of prevention and remediation were found to be significant. Apart from environmental legacies, the mine also left a number of socioeconomic legacies including limited access to non-polluted water that results in San Sebastian residents devoting a high proportion of their income in obtaining water, lost opportunities due to the cessation of mining, uncertain land tenure situation, and increasing growth of ASGM activities that exacerbate already existing environmental pollution due to use of mercury. The study also found that the state's capacity to ensure compliance with the law is very weak and that in many important respects the country's current legal framework does not meet world's best practice when it comes to mine closure requirements.

The findings are important because they demonstrate that the lack of closure planning can lead to private operators socializing the costs of pollution. The study also shows that the lack of state capacity may result in extractive projects becoming socioeconomic liabilities in the long term.

Introduction

This book assesses a number of legacies left by the San Sebastian Gold Mine in eastern El Salvador. The primary objective of the study is to analyze the mine's legacies on the inhabitants of the district of San Sebastian and to provide strategies for awareness, prevention, and remediation to those concerned with mine closure at the local, national, and international levels. This objective is also relevant to those interested in managing limited water resources in communities located near extractive industry sites where pollution may be a problem.

To be able to account for mine legacies in San Sebastian, the analysis took into account legal, environmental, and socioeconomic factors specific to the site as well as its wider context. In terms of water, for instance, San Sebastian shares many of the same difficulties as the rest of the country: organic, industrial, and agro-chemical wastes contaminate most of the surface and groundwater.[1] It is necessary then to see legacies such as acid mine drainage (AMD), not as unique problems but as additional contamination that aggravates and contributes to existing problems. However, in other instances, legacies such as the growth of artisanal mining represent a quite unique phenomenon within the Salvadorean context.

---

[1]The country's ministry of environment (MARN) concluded that of 124 sampling sites used in a 2010 report, none presents as "excellent" or "good" water, only 11% of the sites meet the national standard of fitness for the use of raw water for drinking, only 12% meet the fitness for the use for irrigation, and only the equivalent of 3% of the tested waters meet the WHO standard for recreational activities involving human contact (MARN 2010: 35). And the quality of groundwater is not better. In 2005, a report by the NGO Unidad Ecologica Salvadoreña concludes that pollution levels ranging from "moderate" to "high" (Ibarra et al. 2005: 34). Despite the current and past administration's efforts at improving the quality of surface water, many reports alert us to the dire situation that the country finds itself in (Moreno 2005; Kingsbury 2014; Karunananthan and Spronk 2015). These and other studies carried out in the late 1990s by Salvadorean universities and NGOs are the result of vigorous campaigning by civil society organizations such as the Foro del Agua, Red de Agua y Saneamiento de El Salvador, Alianza Ambiental, Alianza por el Agua, and La Mesa Nacional en Contra de la Mineria Metalica, to improve the quality of water available in the country but also to consider water as a fundamental human right that should not be threatened by short-term economic needs.

A secondary objective of the book is to make a contribution to the growing academic literature on mine legacies. Work carried out in the last 20 years by academics, industry bodies, and government agencies has highlighted the need to minimize negative legacies by planning and financing for closure during the proposal phase of any mining operation.

The result from this growing awareness in Latin America and internationally is that, as is the case in many other industries, there is no agreement on a single definition of what a negative legacy is and what kind of remediation actions this may entail. The lack of agreement is mostly based on the representation of narrow economic and sectoral interests and use of conflicting narratives. Governments and international financial organizations along with the mining industry tend to use a more restrictive and narrow definition of legacy than those used by civil society organizations, which tend to use a definition that is broader in scope. However, one aspect they all seem to agree with is that a lack of definition of what is a negative mine legacy has often resulted in a lack of action, unclear and weak legislation, and social conflict when it comes to mine closure, remediation, and relinquishment. In jurisdictions such as the USA and El Salvador, the lack of definition of what constitutes mine closure and who is responsible for remediation of sites has often ended in long legal battles. But even in jurisdictions with strong mining legislation (such as Australia), there is a tendency to treat legacies as the result of mining that occurred long ago and whose legal owner no longer exists. This view tends to take the focus away from sites with current leases and titles where negative legacies are developing. As Pepper et al. (2014) explain: "while it can be important to distinguish between abandoned and orphan mines in terms of responsibility, liability, solutions and management response; to focus only on abandoned mines is to ignore the problem that exists in existing leases and titles." Therefore, in order to avoid this narrow view and in order to ensure the greatest contribution to the ongoing debates on mining legacies, this book relies on Whitbread-Abrutat's definition of mine legacies that says:

> [Legacies are] the impacts of a mine that continue to negatively affect the environment or associated communities…[and includes] abandoned sites 'where the owner is known, but for some reason, is unable or unwilling to take the necessary remedial action' and orphaned sites 'where the legal owner cannot be traced' (2008: 3).

As we will see below, the San Sebastian Gold Mine clearly fits with the above definition because it is an abandoned mine that is currently affecting the surrounding environment and its people, but its previous owner, Commerce Group Corporation (CGC), has so far shown unwillingness to take any remedial action. Building on that definition, the book focuses on quantifying legacies, finding appropriate remediation strategies and analyzing regulatory failures in the San Sebastian case so that state authorities in El Salvador and any other jurisdictions can use the findings from this study to minimize negative mine legacies in the future.

The content of this book is presented in the following manner: Chapter 1 describes the methodology used in the study. Chapter 2 includes a discussion on closure regulations as the theoretical perspective used in this book. Chapter 3 provides some social and historical remarks about mining in San Sebastian and an assessment of the mine's current legal status. Chapter 4 presents the environmental and socioeconomic legacies of the mine and the findings are presented in 4 different sections. The first section surveys the environmental legacy of AMD contamination. The second section examines how this form of contamination affects people's water use and health. The third section looks at economic impacts, focusing on government payments and land tenure. The last section of this chapter also examines economic impacts but this time focusing on the growth of artisanal mining since CGC abandoned San Sebastian. Chapter 5 discusses a number of reforms needed for the country's legal framework and some ideas on remediation to avoid further damage to the district and its inhabitants. This chapter also discusses a number of prevention and awareness strategies in order to avoid problems when implementing successful closure of mining projects.

References

Ibarra ÁM, Jarquín UC, Rivera FJ (2005) Hacia la gestión sustentable del agua en El Salvador. Propuestas Básicas para Elaborar una Política Hídrica Nacional. San Salvador, Unidad Ecológica Salvadoreña (UNES)/Caritas

Karunananthan M, Spronk S (2015) Water at the heart of El Salvador's struggle against neoliberalism. Blue Planet Project, Ottawa

Kingsbury D (2014) Gold, water and the struggle for basic rights in El Salvador. Oxfam, Melbourne

MARN (2010) Informe de Calidad de Agua de los ríos de El Salvador. San Salvador. Ministerio de Ambiente y Recursos Naturales

Moreno R (2005) El marco jurídico para la privatización del agua en El Salvador. Berlin, Brot für die Welt

Pepper M, Roche CP, Mudd GM (2014) Mining legacies—understanding life-of-mine across time and space. Paper presented at the Life-Of-Mine Conference, Brisbane

Whitbread-Abrutat P (2008) Mining legacy survey, informing the background paper. Post-Mining Alliance, Eden Project, UK

Chapter 1
Methodology

Keywords Rapid Rural Appraisal · Participatory Rural Appraisal · CARITAS · Iterative inquiry process · Interdisciplinary approach · CEICOM

1.1 Research Technique

Given the foci of the research, the process of enquiry in this book is iterative or moving from the general to the specific and vice versa. Sometimes national and international trends and statistics are reflected and/or validated in local realities. At other times the reverse occurs: a local observation requires larger trends and explanations in order to make sense. The most appropriate technique for this iterative process is the Rapid Rural Appraisal (RRA). There are many reasons for using this technique in assessing mining legacies:

- Versatility. Even though RRAs were originally used for rural farm and agricultural projects, these studies have proven to be very versatile and researchers have also used them for all kinds of rural development projects including, but not limited to, health, nutrition, emergencies and disasters, non-formal education, agroforestry and community improvement.
- Cost effectiveness. RRAs are carried out quickly and with minimal human resources but make use of a mixture of quantitative and qualitative methodologies without significantly affecting the reliability of the study (Teddlie and Tashakkori 2003). A total of 60 days spread over a period of 15 months were employed in this study and included two 10 day field trips to El Salvador (with a 3 day stay in San Sebastian) in October 2014 and January 2016 and 2 subsequent site visits by a local research assistant. Throughout the study the author maintained remote contact with some of the key informants in San Sebastian in order to corroborate information that emerged after the visit.
- Reduced urban and expert bias. Chambers refers to RRAs as "fairly-quick and-fairly-clean" studies, as opposed to "quick-and-dirty" and "long-and-dirty" studies in Crawford (1997). Chambers' use the term "dirty" refers to the urban and conceptual biases that many "experts" display when doing socioeconomic

© The Author(s) 2017
V. Pacheco Cueva, *An Assessment of Mine Legacies and How to Prevent Them*,
SpringerBriefs in Environmental Science, DOI 10.1007/978-3-319-53976-8_1

research in rural areas, a factor that ultimately leads to incorrect or skewed results. RRAs, on the other hand, reduce urban and expert bias by relying heavily on knowledge generated from key informants (village elders, community leaders, field-based government officials)[1] and the active participation of the community in the research process. RRAs also use triangulation, or "using more than one technique/source of information to compare and complement information from different sources or gathered in different ways" (Crawford 1997). Therefore, in order to keep the results of research fairly quick and clean and minimize urban and expert bias the study was carried out both in situ and remotely and includes the participation of key informants such as the inhabitants of the affected areas, government officials and local NGOs working in areas of community development, water and public health. In order to triangulate the information provided by key informants the research makes reference to statistical data from government ministries, the Central Bank, NGOs and previous studies on subjects such as environmental pollution, mining, human development and water consumption. The research also takes into account relevant national and international academic literature on the subject of mine legacies in both English and Spanish.

1.2   Data Collection

Relying on various sources of information means that RRAs can generate large amounts of data within a short period of time. This also means that, often, it can become difficult to separate the ways in which data were collected. Still, three forms of data collection and calculation methods are distinguishable in this study: in situ or field observations; key informant interviews; and analysis of secondary sources. The in situ or field observations are qualitative in nature as well as the interviews with key informants. Secondary sources consisted mostly of quantitative data obtained from government agencies, other studies carried out in the area (such as those by the Centro de Investigación sobre Inversión y Comercio-CEICOM) and expert opinion. Quantitative data from secondary sources was supplemented with queries to government agencies and local NGOs. The study also made use of results from a CARITAS Participatory Rural Appraisal (PRA) carried out earlier in 2008 at La Presa hamlet, San Sebastian. Additional information about the socio-economic environment of the municipality of Santa Rosa de Lima and the department of La Union came from a variety of sources, including the government's statistical agency (DIGESTYC), the central government's water authority (ANDA), the Ministry of

---

[1]The research relied on the cooperation of two local NGOs: ASPRODE and CRIPDES. These NGOs and the communities of San Sebastian have a long history of collaboration meaning that institutional knowledge of the area is strong. Furthermore the author hails from the eastern part of the country and his family has intimate knowledge of the territory where the mine is located.

Health and the Central Bank. Also, on several occasions under the auspices of the new law on information access for citizens, the author made use of the Office of Information and Response (OIR) at the Ministry of Economy (Directorate of Hydrocarbons and Mines) and the Ministry of Finance. Interviews were also carried out with officials from the Ministry of Environment and Natural Resources, the Ministry of Agriculture and Livestock and the Hospital in Santa Rosa de Lima.

The in situ inspection, interviews with key informants and analysis of secondary sources provided information to suggest that there are four main legacies left by the San Sebastian mine:

- The AMD legacy impacting surface and ground water.
- The socio-economic impacts of AMD.
- The land tenure legacy.
- The growth of artisanal and small scale mining (ASGM) legacy.

For reasons of space and simplicity the methodology used for each one of these legacies is explained in Chap. 3.

References

Crawford IM (1997) Marketing research and information systems (Marketing and Agribusiness Texts—4). Food and Agriculture Organization, Rome
Teddlie C, Tashakkori A (2003) Major issues and controversies in the use of mixed methods in the social and behavioral sciences. In: Tashakkori A, Teddlie C (eds) Handbook of mixed methods in social and behavioral research. Sage, Thousand Oaks

Chapter 2
Knowledge About Mine Legacies, International Best Practice Standards and Mine Closure Regulation in the USA and El Salvador

Keywords Mine legacies · El Salvador mining law · Berlin guidelines · CAMMA · ASGMI · OCMAL · Equator Principles · EITI · International Cyanide Management Code · Minamata Convention · Conflict-Free Sourcing Initiative

2.1 What We Know About Mine Legacies

Increasing awareness of the long-term negative impacts of mining amongst civil society and academic actors, industry groups and governments has spurred a number of publications, projects and events that seek to define, classify, analyze, assess and document negative mine legacies. Table 2.1 provides a selection of these activities.

Similar publications and projects emanating from Latin America demonstrate heightened awareness of mining legacies in the region and reflect international developments. Some notable examples of work carried out in this field include the 2001 proceedings of the Latin American Mining Minister's Annual Conference (CAMMA), the Iberian-American Association of Mining and Geological Services' (ASGMI) definition of environmental liabilities[1] and the ongoing Latin American Observatory of Mining Conflicts' (OCMAL) own database on mine legacies.[2] Academic works include those of Villas Bôas and Barreto (2000), Guerrero-Almeida et al. (2014) and Salazar Pérez and Montero Peña (2014). In addition to this work,

---

[1]According to *ASGMI*, mining environmental liabilities include facilities, buildings, areas affected by discharges, deposits of mining waste, disturbed channels, workshops, machinery areas, ore deposits in currently abandoned or paralyzed mines that constitute a permanent potential risk to the health and safety of the population, biodiversity and the environment (ASGMI 2010: 2).
[2]OCMAL's database includes descriptions of the main environmental liabilities per mining site. See http://www.conflictosmineros.net/?Itemid=44. OCMAL's other work of interest is its maps on cyanide use and mining conflict. See http://www.conflictosmineros.net/images/stories/MAPA-CIANURO-AL.jpg and http://mapa.conflictosmineros.net/ocmal_db/.

© The Author(s) 2017
V. Pacheco Cueva, *An Assessment of Mine Legacies and How to Prevent Them*,
SpringerBriefs in Environmental Science, DOI 10.1007/978-3-319-53976-8_2

Table 2.1 International activities in relation to mine legacies and closure since 1995

Year	Author/organiser	Type	Description	Focus
1995	Various US Federal agencies	I	Federal Mining Dialogue	USA
2000	MiningWatch Canada	P	*Mining's toxic orphans*	Canada
2000	Veiga et al.	P	*Filling the Void: The Changing Face of Mine Reclamation in the Americas.*	International
2001	The Eden Project/Post mining Alliance	IP/I	Post-mining tourist attraction and educational facility opens	UK
2001	Cochilco/UNEP	P	*Abandoned mines—problems, issues and policy challenges for decision makers*	International
2001	Canadian Government	W	Workshop on orphaned/abandoned mines	Canada
2002	WB/IFC	W/P	*It's not over when its over: mine closure around the world*	International
2002	Cochilco	P	*Research on mine closure policy*	International
2002	South African Government	L	Introduction of the Minerals and Petroleum Resources Development Act	South Africa
2002	NOAMI	I	National Orphan/Abandoned Mines Initiative established	Canada
2003	ACMER	W/P	Management and remediation of abandoned mines	Australia
2005	MCMPR	I	Formation of the Abandoned Mines Working Group (AMWG)	Australia
2005	Clark and Clark	P	*An International Overview of Legal Frameworks for Mine Closure*	International
2006	NOAMI	W	Orphaned and abandoned mines: a workshop to explore best practices	Canada
2006	Macdonald et al.	P	*Integrated Closure Planning Review: Literature Review*	International
2008	IUCN-ICMM	W/P	Roundtable on restoration of legacy sites	International
2008	NOAMI	W	Workshop to explore perspectives on risk assessment of orphaned and abandoned mines	Canada
2013	AusIMM	P	Abandoned mine policy statement and annexure	Australia
2016	Australian Centre for Geomechanics	C	Mine Closure Annual Conference	International

P Publication; *I* Initiative; *W* Workshop; *C* Conference; *L* Legislation; *B* Infrastructure project
Table modified from Pepper et al. (2014: 452)

the Mine Closure Conference 2017 and the newly established Mining for Closure International Congress (organized by the Sustainable Minerals Institute and Gecamin) will take place in Peru and Chile, respectively and both events deal with mining legacies in Latin America and beyond.

The overall results from all this activity is that some conceptual advances have been achieved in the last 20 years and this, in turn, has informed policy and practice within companies and regulatory schemes within governments. For instance, in one of its reports, the Salvadorean ministry for the economy (MINEC) reflects international thinking when it declares that: "at present, most countries in the world consider mine rehabilitation as part of mine exploitation in their mining codes" (MINEC 2011: 53). However, this insight has not translated into improvements to the current mine legislation and practice in El Salvador. A look at mining closure regulation in the USA (where CGC is registered) and El Salvador reveals a number of gaps when compared to international best practice.

2.2 International Best Practice on Mining Closure and Reduction of Mine Legacies

At the outset it must be said that there are no international covenants or conventions that regulate mining closure and its legacies. Some industrial processes, such as those relying on cyanide and mercury are covered by international conventions (International Cyanide Management Code and the Minamata Convention). Other activities such as the sourcing of some non-renewables are covered by industry led certification programs (The Conflict-Free Sourcing Initiative) and another program emphasizes transparency over the proceeds from mining (the Extractive Industries Transparency Initiative). The closest form of regulation at the international level comes from mining industry adherence to 'international best practice' standards, guidelines and voluntary schemes. These practices are disseminated and sometimes verified by international industry bodies and international organizations but generally there are no enforcement mechanisms, only reputational risks. Some voluntary schemes (such as the Equator Principles) include financing arrangements that are tied to social and environmental criteria that can be applied before government approval of a natural resource extraction project. According to Bastida (2005), some of these standards may have been influenced by the Berlin Guidelines which were developed at the Round Table on Mining and the Environment held in Berlin in 1991. The Guidelines focus on mine site stability and the socio-economic impacts of closure, especially in regards to future uses of lands surrounding mine sites. International financial institutions such as the International Finance Corporation (IFC) and the World Bank (WB) followed suit by developing standards and guidelines that borrowers must comply with before obtaining finance.[3] Some of the WB guidelines address the financial feasibility of mining projects when the following elements are taken into account:

- The costs of closure,
- Premature or temporary closure,
- Post closure and
- The type of institution that would provide a financial guarantee for this phase of the project.

Just as international financial institutions have created standards for closure, the international mining industry, represented through the ICMM, has also developed its own set of guidelines for closure. The ICMM's Planning for Integrated Mining Closure Toolkit (2008) has very detailed steps based on the principle that planning for closure begins at the exploration phase. In other words the Toolkit assists mining companies follow international best practice by changing the way closure is perceived right from the start of operations and includes financial, environmental, social and economic aspects of mine closure.

2.3 Mine Closure Regulation in the USA

A comparison between US Federal law and international best practice standards shows that with the possible exception of coal mining (Slezak and Robertson 2016), mine closure regulation in the USA is relatively weak. After some strengthening of the 3809 Mining Rules in 2000, under the Clinton administration, the Bush administration basically repealed all of these improvements. As a result current Federal mine closure regulation in the USA relies heavily on financial assurance without a correspondingly clear definition of mine rehabilitation. Kuipers explains the problem thus:

> The Clinton rules better protected the environment in general—and surface water and groundwater resources in particular—by defining standards for mine operation, reclamation and closure. It also significantly improved federal financial assurance by: requiring approved financial assurance as a precondition for mining; eliminating new corporate self-guarantees; authorizing the establishment of trust funds to address long-term water treatment costs; and codifying authorization of incremental financial assurance release...In October of 2001, the Bush administration rolled back almost the entire rule, including those parts defining reclamation and closure... Although the financial assurance section of the rule was kept verbatim, [this] is only as strong as the definition of mine reclamation. The Bush administration essentially eliminated the definition of mine reclamation—leaving the

---

[3]This includes the IFC's Policy and Performance Standards on Environmental and Social Sustainability (originally published in 2006 but revised in 2012), the Environmental, Health and Safety Guidelines for Mining (2007) and the WB's Towards Sustainable Decommissioning and Closure of Oil Fields and Mines: A toolkit to Assist Government Agencies, Version 3 (2010) and the Guidance Notes for the Implementation of Financial Surety for Mine Closure (2008).

estimator of the financial assurance (usually the mining company) to define it on a site-by-site basis (Kuipers 2003: 25).

The shortcomings outlined by Kuipers may be compensated by the fact that the 3809 Mining Rules are supplemented by interrelated regulations at the Federal and State levels. Some of these supplementary regulations, such as the Clear Water Act, the Surface Mining Control and Reclamation Act and the Toxic Substances Act, amongst others, provide an additional layer of protection against unplanned mine closure. Still, according to Kuipers, taxpayers in the USA "are potentially liable for between $1 billion to more than $12 billion in clean-up costs for hardrock mining sites" (2003: 2).

From the foregoing it can be said that there are two problems with the US mining law as it relates to international best practice. The first is that the law is heavily reliant on strong financial assurance without a correspondingly strong definition of mine closure. It is difficult then to see how this law can be used as a guide to improve mine closure law in El Salvador.[4] The second problem has to do with adaptability and/or transferability of certain legislation. As was said earlier, Federal laws are supplemented by legislation at both the Federal and State levels. El Salvador's legal system on the other hand is unitary, i.e. despite the fact that municipalities can enact laws these are usually subsumed under national level laws. This means that there is no complementarity of the two levels as is the case in a federal system. The lack of complementarity means that a single set of regulations at the national level must contain all of the safeguards needed to protect the environment. US laws are therefore not a good guide to improve El Salvador's mine closure regulation.

2.4   Mine Closure Regulation in El Salvador

The Salvadorean mining code was originally introduced in 1876 and modified in 1881 and 1884. A second mining code was introduced in 1922 and modified in 1953, 1957, 1960, 1986 and 1989 (Vildosola 1999: 220). The current law was introduced in 1996 (*Ley de Mineria*), with further revisions in 2001 and 2003 and is accompanied by a respective set of by-laws (*Reglamento Ley de Mineria*). The law does not explicitly lay out a procedure or requirement for mine closure. Instead it is the Environmental Act of 1998 (*Ley del Medio Ambiente*) and the accompanying by-laws (*Reglamento General de La Ley de Medio Ambiente*) that determine the requirements for closure. The by-laws require companies to establish an environmental management program (*Programa de Manejo Ambiental*) as part of the

---

[4]This is also significant in terms of compliance. It is reasonably to assume that CGC and its managing personnel, despite being familiar with the US Federal legal requirements for closure (along with those from Wisconsin/Nevada, where CGC is registered), proceeded to submit to the less stringent Salvadorean mining legal requirements in order to save money.

Environmental Impact Study (EIS). An EIS is a precondition for an environmental permit, which is, in turn, a requisite for a mining permit. The environmental management program must include the following components:

a. Determination, prioritization and quantification of prevention measures, mitigation and compensation of environmental impacts and determination of necessary investment;
b. Monitoring;
c. Closure of operations and rehabilitation; and
d. A study of risk and environmental management, in cases where it is necessary.

More specifically, in relation to closure, Article 27 of the by-law estipulates:

> The closure of operations and rehabilitation component, *when these occur*, will aim to identify and determine the environmental measures to be adopted and implemented by the holder, during *or after* the closure of operations, *as appropriate*, as well as those required to restore the damage caused during the operational stage. This component will contain the description, location, investment estimate and schedule for implementing the measures (my emphasis).

The responsibility for enforcing mine closure therefore rests with the Ministry for the Environment and Natural Resources (MARN) through Articles 24 to 29. There are other articles in the mining and environmental regulations that, when taken in conjunction, could be interpreted as requiring companies to address and/or compensate for any environmental, social and economic impacts caused by the closure of a mine.

Despite the improvements made to the latest version of El Salvador's mining law, a lot remains to be done to fulfill the requirements of international best practice. As will be shown in the subsequent chapters Salvadorean mining law is not only insufficient in regards to the financial aspects of closure as set by the World Bank and the IFC but also contains mechanisms that allow companies to subvert planning for closure as set by the ICMM Closure Toolkit. The government is also aware of other gaps related to the protection of water bodies during the operative and closure phases of mining. A comprehensive assessment of the regulatory framework carried out by MINEC in 2011 has the following assertions in this regard:

> The only explicit legal body that determines priority uses of water in El Salvador is Presidential Decree 51 (1987). This is a positive instrument but it is difficult to enforce and therefore not effective from a legal point of view. There are no express measures for the protection of receptor water bodies, nor aspects related to territorial vulnerabilities. There are also technical details that have not yet been regulated, such as the handling and disposal of wastes, the processing of sludge from mining activities, wastewater reuse, or management of persistent organic pollutants (POPs) not intentionally generated by mining activity or operation (MINEC 2011: 56).

This situation minimizes prevention and increases the likelihood of environmental damage as a result of mining operations and lack of proper closure and, as will be shown later in the book, it is best exemplified by continued AMD contamination of the water in and around the San Sebastian River.

This section has compared international best practices standards and relevant literature on mine closure with current legislation in the USA and El Salvador. From these comparisons it is clear that mining laws in the USA are behind, in many respects, to what is considered international best practice and at the same time are not easily transferable to the Salvadorean mining system. It is also clear that a number of gaps exists in the Salvadorean mining law and that a thorough reform of the existing law and associated regulations in the Environmental Law are necessary if negative legacies are to be minimized or avoided in the future.[5]

References

ASGMI (2010) Pasivos ambientales mineros. Manual para el inventario de minas abandonadas o paralizadas. Barquisimento. Asociación de Servicios de Geología y Minería Iberoamericanos

Avalos Sánchez MG, Castillo Figueroa CA (2008) "Concesión de explotación de recursos yacentes en el subsuelo de la República de El Salvador". Trabajo de investigacion para obtener el grado y titulo de: licenciado en ciencias juridicas. Escuela de Ciencias Juridicas, Facultad de Jurisprudencia y Ciencias Sociales, San Salvador, Universidad de El Salvador

Bastida E (2005) Interfaces between international standards, national legislations and corporate self-regulation regarding mine closure: the experience of Argentina, Bolivia and Chile. Scotland, Unpublished LLM dissertation, University of Dundee

Guerrero-Almeida D, Chacón-Pérez Y, Fonseca-Hernández D, Court-Potrillé M (2014) Metodología para la ejecución de un cierre de minas sustentable. Minería Geol 30(3):85–103 Julio-Septiembre

ICMM (2008) Planning for integrated mining closure toolkit. International Council on Mining and Metals, London

IFC (2007) Environmental, health and safety guidelines for mining. International Finance Corporation/World Bank, Washington

Kuipers J (2003) Putting a price on pollution: financial assurance for mine reclamation and closure. MPC Issue Paper No. 4. Washington, Mineral Policy Center

Maldonado Tejada GM, Martínez López JY, Murcia Perdomo NM (2010) El impacto que puede generar la explotación minera en el derecho fundamental a la vida de la población salvadoreña. Trabajo de investigaciòn para obtener el grado de: licenciatura en ciencias juridicas. San Salvador. Escuela de Ciencias Juridicas, Facultad de Jurisprudencia y Ciencias Sociales, Universidad De El Salvador

MINEC (2011) Evaluación Ambiental Estratégica (EAE) del sector minero metálico de El Salvador. Servicios de consultoría, Concurso publico internacional Nº CPI—02/AECID/2010. San Salvador, Unidad de Cooperación Externa

Pepper M, Roche CP, Mudd GM (2014) Mining legacies—understanding life-of-mine across time and space. Paper presented at the Life-Of-Mine conference, Brisbane (July)

Rivas Cruz JJ, Palencia Perez OH (2010) La responsabilidad social corporativa y la vulneración a los derechos humanos en El Salvador: el caso de dos empresas multinacionales mineras. Trabajo de investigacion para obtener el grado de: Licenciado en Ciencias Juridicas. San

---

[5]There are other issues of consideration that are beyond the scope of this study. Many of these concerns have been raised through legal studies. See for instance Maldonado Tejada et al. (2010), Rivas Cruz and Palencia Perez (2010), Avalos Sánchez and Castillo Figueroa (2008).

Salvador. Escuela de Ciencias Jurídicas, Facultad de Jurisprudencia y Ciencias Sociales, Universidad De El Salvador

Salazar Pérez Y, Montero Peña J (2014) La planificación del cierre de minas como parte de la sustentabilidad en la minería. Observatorio de la Economía Latinoamericana, N° 199. Retrieved May 07, 2016 from http://www.eumed.net/cursecon/ecolat/cu/2014/minas.html

Slezak M, Robertson J (2016) Full of holes: why Australia's mining boom will leave permanent scars. The Guardian Australia. Tuesday July 19

Vildosola FJ (1999) El Dominio minero y el sistema concesional en América Latina y El Caribe. Santiago de Chile, CEPAL-OLAMI

Villas Bôas R, Barreto ML (2000) Cierre de Minas: experiencias en Iberoamérica. Río de Janeiro, CYTED/MAAC/UNIDO

Chapter 3
Social and Historical Remarks About Mining in San Sebastian and Assessment of the Mine's Current Legal Status

Keywords Commerce Group Corporation · El Salvador gold mining history · International Centre for Settlement of Investment Disputes (ICSID) · Abandoned mines · Orphaned mines · San Sebastian Mine · Santa Rosa de Lima

After years of civil war (1978–1992), El Salvador remains a fragile country both politically and economically. With a geographical territory of 21,000 km², a population of 6.2 million and scare water resources, the country struggles to provide a decent livelihood for all of its citizens and ranks only 116 out of 187 countries according to the UN's 2015 Human Development Index. Since the end of the civil war in 1992, successive governments of this small Central American nation have sought to industrialize the country by opening up the economy and encouraging foreign investment. Since the early 2000 as many as 10 transnational mining companies started prospecting for gold and other minerals around the mountainous and forested northern parts of the country. In response to mining industry interest in these resources, civil society organizations raised concerns over food sovereignty and environmental degradation as well as the social division that these projects may cause or aggravate. As a result of this pressure, the government stopped issuing mining exploration and exploitation permits at the national level in early 2008 and since, two successive governments have upheld the moratorium. Even though this measure has effectively halted formal metallic mining in San Sebastian, artisanal mining continued to operate informally soon after CGC stopped operations. Artisanal miners take advantage of the close relationship between the mine and the district's capital, Santa Rosa de Lima. The town, located only 5 km from the mine, is a commercial centre that thrives on buying and selling smuggled goods from nearby Honduras and Nicaragua. The town is also regarded as a market for gold that attracts costumers from the neighbouring countries and from elsewhere in El Salvador (MINEC 2011: 25). Gold buyers and jewelry shops can be found everywhere in the town's main streets and this means that artisanal miners have nearby outlets to sell their production.

© The Author(s) 2017
V. Pacheco Cueva, *An Assessment of Mine Legacies and How to Prevent Them*,
SpringerBriefs in Environmental Science, DOI 10.1007/978-3-319-53976-8_3

3.1   Growth and Decline of the Mine

Despite the miner's ubiquitous economic impact on the town, their presence pales in comparison to what occurred earlier in the history of the district. According to MINEC (2015), mining in San Sebastian goes back to 1904 and became more intense between 1935 and 1953 with a total output of 180,000 oz of gold produced during that period. The United Nations' 1969 Mineral Survey Report stated that "unquestionably the San Sebastian deposit was the jewel of the El Salvador mining industry and one of the most prolific gold mines in Central America" (quoted in SEC, n.d.: 18).[1] According to El Salvador's National Registry, in 1960 the owner of the mine was Minerales San Sebastian, S.A. de C.V. (Minsanse, which in turn was owned by 90 workers from the mine, whom had 4 shares each in the company). Subsequently in 1969, the mine became the property of San Sebastián Minerals and continued operations until 1978, when it closed as a result of the civil war in El Salvador and labour strife within the company. The mine resumed operations in 1985 and in 1987, Commerce Group Corporation (CGC) and San Sebastian Gold Mines (SSGM) entered into a joint venture registered in Wisconsin, USA, to explore, develop, mine and produce precious metals in El Salvador.[2] The 1990s become another 'golden age' for the company: in the four years from 1995 to 1999, the 200 tonnes per day capacity of the Joint Venture's San Cristobal Mill and Plant (SCMP)

---

[1]MINEC (2015) warns that official statistics about mining are not very reliable during that period so the mine's output should be treated with caution.

[2]This is officially known as the "Commerce/Sanseb Joint Venture". In this venture CGC owns 82.5% of the authorized and issued stock of SSGM.

was able to produce 13,305 oz of gold and 4667 oz of silver (ICSID 2009: 3). By 2000 CGC suspended its gold production until "able to procure the funds it requires to rehabilitate, retrofit, overhaul, and expand its SCMP and/or when it has funds to commence an open-pit, heap-leach operation at the SSGM site" (SEC 2004).

Despite attempts at increasing production and renewing mining concessions in the early 2000s, the company faced regulatory difficulties in 2006 and was neither able to regain its licenses nor carry out mining operations in the country. As a result of losing its environmental permits, CGC decided to take the government of El Salvador to court to recoup investment costs in the national courts. When the court ruled against the company, its managers decided to take the matter to the International Centre for Settlement of Investment Disputes. The case between CGC versus The Republic of El Salvador (2009–2013) is well documented and there is no need to repeat it at length in this book. The following is a simplified chronology of events as recorded in the proceedings of ICSID Case No. ARB/09/17 (2009: 18–19).

- The Commerce/Sanseb Joint Venture received an exploitation concession from the Government of El Salvador for the San Sebastian Gold Mine on 23 July 1987. At this time, [the Joint Venture] and Minsanse entered into an agreement to lease 305-acres at the San Sebastian Gold Mine (the "Minsanse Agreement").
- Later, in 1993, [the Joint Venture] acquired two additional properties, the El Modesto Mine and the San Cristóbal Mill and Plant.
- On 18 August 2002, [the Joint Venture] met with the El Salvadoran Minister of Economy and the Department of Hydrocarbons and Mines to cancel their exploitation concession license for the San Sebastian Gold Mine in exchange for another exploitation license, to last for 20–30 years.
- In order to mine and process gold ore at the San Sebastian Mine and San Cristóbal Mill and Plant, [the Joint Venture] received environmental permits from the El Salvador Ministry of Environment and Natural Resources (the "MARN") on 20 October 2002 and 15 October 2002, respectively, renewed for a 3-year period as of 4 January 2006.
- In addition, El Salvador granted [the Joint Venture] two further exploration licenses, namely: (i) on 3 March 2003, encompassing the San Sebastian Mine and adjoining areas (the "New San Sebastian Exploration License"); and (ii) on 25 May 2004, encompassing eight former gold and silver mines (the "Nueva Esparta Exploration License").
- On 13 September 2006, MARN revoked the environmental permits of the San Sebastian Gold Mine and the San Cristóbal Plant and Mine, thereby effectively terminating [the Joint Venture's] right to mine and process gold and silver.
- In response, on 6 December 2006, counsel for Commerce and SanSeb filed two petitions with El Salvador's Court of Administrative Litigation of the Supreme Court of Justice, one for each affected mine, seeking a review of the Ministry of the Environment's revocation of the environmental permits and their reinstatement.

- In the interim, over the course of 2006 and 2007, Commerce/Sanseb applied to MARN for an environmental permit for the New San Sebastian Exploration License and the Nueva Esparta exploration license, and then to [the] Ministry of Economy for the extension of the exploration licenses. The requested environmental permits were not granted, and on 28 October 2008, El Salvador's Ministry of Economy denied Commerce/Sanseb's application citing Commerce/Sanseb's failure to secure an environment permit.
- In July 2009 the Joint Venture applied to the ICSID to rule on the case and therefor waive their right to continue proceedings in the national courts.
- On 29 April 2010, El Salvador's Court of Administrative Litigation of the Supreme Court of Justice notified its decisions of 18 March 2010 (Case No. 308-2006) and 28 April 2010 (Case No. 309-2006) with respect to these two complaints [The Company lost the case.].
- On August 28, 2013, the ad hoc Committee of the ICSID issued its Order of the Committee Discontinuing the Proceeding and Decision on Costs. In summary, it was ordered that the annulment proceeding be discontinued, that the Company and the Government of El Salvador each bear its own costs and legal expenses and that the expenses and fees of the Members of the ad hoc Committee and the charges for the use of the facilities at ICSID shall be borne by the Company. A copy of the Order dispatched to the parties on August 28, 2013, [was] posted on the Company's website: http://www.commercegroupcorp.com/.

3.2   Current Status of the Mine Owner

According to Whitbread-Abrutat the difference between abandoned and orphaned mines is that the former are sites 'where the owner is known, but for some reason, is unable or unwilling to take the necessary remedial action' while the latter are sites 'where the legal owner cannot be traced' (2008: 3). Beyond the academic exercise of classifying the nature of mining legacies, these categories are important because in legal and administrative disputes they help attribute responsibility for negative legacies. In order to categorise the San Sebastian mine as either an abandoned or an orphan site it was therefore important to establish the current status of the mine owner.

It seems that after the ICSID determination against CGC in 2013, the Joint Venture went dormant and its physical and administrative presence in El Salvador has been reduced to null. For instance, the CNR database has a listing of land titles but information related to the CGC lease has not been updated since 2011 and it was therefore not possible to establish the current legal situation of the lease. The lease in question is depicted in the following map:

Misanse and El Paraiso Map
– San Sebastian Gold Mine Lease (SSGM)
El Salvador, Central America

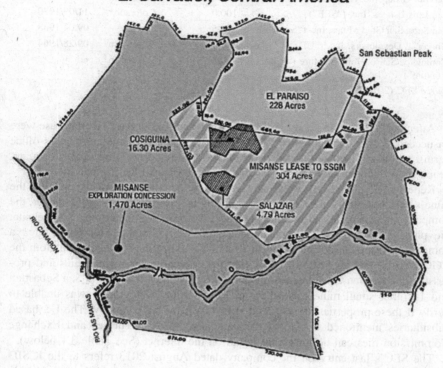

San Sebastian Peak

EL PARAISO
228 Acres

COSIGUINA
16.30 Acres

MISANSE LEASE TO SSGM
304 Acres

MISANSE
EXPLORATION CONCESSION
1,470 Acres

SALAZAR
4.79 Acres

RIO CAMARON

RIO LAS MARIAS

RIO SANTA ROSA

Three Areas	Acres
El Paraiso Property	228
Misanse Exploration Concession	1,166
Misanse Property Leased to SSGM	304
Misanse Property Total	1,470

Scale

1 : 20,000 Meters

Source: SEC, n.d: 23

Mineralized Material

L = 1.0 km. or 3,300 ft.
W = 250 mts. or 825 ft.
D = 200 mts. or 660 ft.

Estimated Average Grade: 0.025 ounces of gold per ton

Total tonnage   138,219,230
Total estimated ounces of gold: 3.4 million
Total estimated ounces of silver: 0.4 million

Table 3.1 CGC Subsidiaries as of end 2010

Name	% Ownership	Place	Establishment date
Homespan Realty Co., Inc. ("Homespan")	100.0	Wisconsin	02/12/1959
Ecomm Group Inc. ("Ecomm")	100.0	Wisconsin	06/24/1974
San Luis Estates, Inc. ("SLE")	100.0	Colorado	11/09/1970
San Sebastian Gold Mines, Inc. ("Sanseb")	82.5	Nevada	09/04/1968
Universal Developers, Inc. ("UDI")	100.0	Wisconsin	09/28/1964
Commerce/Sanseb Joint Venture ("Joint Venture")	90.0	Wisconsin	nd

Source SEC (2010: 8)

Attempts at contacting company representatives listed in the CNR database were unsuccesfull. A visit to the city of San Miguel in search of the company's office during a field trip showed that it is no longer at the address it occupied for many years. In Wisconsin, the company is no longer located in its original address.[3]

Repeated telephone calls, faxes and e-mails to the company to help establish the mine's current status were not answered during the time of the study. However, the company still maintains an internet presence. According to the company's website, the physical address in Wisconsin is shared with Circular Marketing, Inc., a company that the current president of CGC has controlling ownership of. Both the CGC website and an SEC file show that the company still owns a mill and processing plant at San Cristobal near the city of San Miguel, the mine at San Sebastian and 13 other small mines elsewhere in El Salvador.[4] The author was unable to verify if these properties were indeed still owned by the company. The US based subsidiaries mentioned in one of the company's 2010 Securities and Exchange Commission files can no longer be found on the internet (See Table 3.1 below).

The SEC's last entry on the company dated August 2013 refers to the ICSID verdict and contains no financial information about the company.[5] It is reasonable to assume that the company subsidiaries have been either sold or went into bankruptcy. This hypothesis is supported by a 2010 SEC file that says that the company's future is uncertain:

> The Company has recurring net losses, negative working capital and negative cash flow from operations, and is dependent upon raising capital to continue operations. The Company's

[3]That address was 6001 N 91st St Ste 2. Milwaukee, WI 53225 Parkway Hills where it rented space from General Lumber & Supply a company in which the long-time Chairman, but now deceased, Mr. Edward L. Machulak had a 55% ownership of.

[4]The smaller mines are called: Bañadero, Carrizal, Copetillo, Grande, La Joya, La Lola, Las Piñas, Montemayor, Oro, Santa Lucia, Tabanco and Modesto (SEC n.d.: 18 and http://www.commercegroupcorp.com/mines.html).

[5]See: https://www.sec.gov/Archives/edgar/data/109757/000010975713000002/0000109757-13-000002.txt.

ability to continue as a going concern is subject to its ability to generate a profit and/or obtain necessary funding from outside sources, including obtaining additional funding from the sale of its securities, increasing sales or obtaining loans and grants from various financial institutions where possible. The consolidated financial statements do not include any adjustments that might result from the outcome of this uncertainty (SEC 2010: 6).

From the foregoing it is clear that the owner of the mine is traceable which means that, following Whitbread-Abrutat's definition, San Sebastian is not an orphaned mine but an abandoned mine. However, from the documents and enquiries that the current study made in the course of 15 months, it was not possible to determine what assests the company or its subsidiaries still have in El Salvador and whether the head office in the USA is operating at all. Therefore, what can be said is that CGC is unable or unwilling to take the necessary remedial action in relation to mine legacies left behind in San Sebastian.

References

ICSID (2009) Commerce Group Corp. and San Sebastian Gold Mines, Inc. v. The Republic of El Salvador. Notice of Arbitration, ICSID Case No. ARB/09/17 19, Washington, International Centre for Settlement of Investment Disputes

MINEC (2011) Evaluación Ambiental Estratégica (EAE) del sector minero metálico de El Salvador. Servicios de consultoría, Concurso publico internacional Nº CPI—02/AECID/2010. San Salvador, Unidad de Cooperación Externa

MINEC (2015) Proyectos de exploracion con mayor avance en la investigacion. Retrieved April 10, 2016 from http://servicios.minec.gob.sv/default.asp?id=67&mnu=50

SEC (2004) Commerce Group Corp, Form 10-K/A. Securities and Exchange Commission, Washington. Retrieved December 21, 2016, from https://www.sec.gov/Archives/edgar/data/109757/000114037705000136/0001140377-05-000136.txt

SEC (2010) Commerce Group Corp, Form 10-Q. Securities and Exchange Commission, Washington. Retrieved April 20, 2016, from http://www.sec.gov/Archives/edgar/data/109757/000010975711000001/dec2010.txt

SEC (n.d.) Commerce Group Corporation Securities and Exchange Commission files. Washington. Retrieved February 26, 2016, from http://google.brand.edgar-online.com/EFX_dll/EDGARpro.dll?FetchFilingHtmlSection1?SectionID=6045022-46820-85862&SessionID=gX-eHSHpJOTcb47

Whitbread-Abrutat P (2008) Mining legacy survey, informing the background paper. UK, Post-Mining Alliance, Eden Project

Chapter 4
Legacies of the San Sebastian Mine

Keywords Acid Mine Drainage (AMD) · Water for human consumption · Improved sanitation facilities · Market price for water · Guillan-Barre syndrome · Chronic kidney disease · Resource curse · Company community expenditure · Artisanal and Small Scale Gold Mining (ASGM) · *Güiriseros* · Mercury discharge · Mercurialism

4.1 AMD as Primary Environmental Legacy in San Sebastian

Acid mine drainage is one of the most common legacies in mining (Pepper et al. 2014: 453). According to the USGS acid mine drainage occurs when:

> [P]yrite, an iron sulfide, is exposed and reacts with air and water to form sulfuric acid and dissolved iron. Some or all of this iron can precipitate to form the red, orange, or yellow sediments in the bottom of streams containing mine drainage. The acid runoff further dissolves heavy metals such as copper, lead and mercury into ground or surface water (SEC, n.d.).

AMD can contaminate surface and ground water and disrupt the growth and reproduction of aquatic plants and animals and these problems can persist for hundreds of years (Evanhoe 2006). In El Salvador, the San Sebastian mine represents the only case of AMD contamination in the country. The history of the area indicates that the mine had been active since the early 1900s. CGC reports that the amount of waste material in the area was relatively large when the company started operating the mine in the late 1960s:

> The Company estimates that it has 14.4 million tons of virgin mineralized material, including the dump waste material. The dump material and stope fill are the by-products of past mining operations. The dump material was mined in the search for higher grades

© The Author(s) 2017
V. Pacheco Cueva, *An Assessment of Mine Legacies and How to Prevent Them*,
SpringerBriefs in Environmental Science, DOI 10.1007/978-3-319-53976-8_4

of gold ore and piled to the side of past excavations as it was considered at that time to be too low a grade of ore to process economically; however, it was reserved for future processing when the price of gold reached a level to process it profitably. The available stope fill was also considered to be too low a grade of mineralized material to process economically, therefore it was primarily used to fill the voids in the underground workings to accommodate the extraction of the higher grade of mineralized material (SEC, n.d.: 18).[1]

Therefore the possibility that AMD discharge was already present when CGC acquired the mine is very high. However, since there is no official record to determine the level of AMD discharge and the volume of water carried by the river at the time, it is very difficult to establish a baseline from which to measure the rate of damage. Without a baseline a possible explanation is that AMD contamination became much more serious as result of particular mining practices used by CGC or as a result of increased production from the time it started operating in the mine.[2]

Regardless of the lack of AMD data from the past, the condition of the soil, water and plants in and around San Sebastian is now well documented. In one of its studies CEICOM reports that at the AMD discharge point the water's pH levels range from 2.70 to 3.01 depending on the season (CEICOM 2012: 6–7). Such high level of acidity dissolves heavy metals into the water and the river sometimes looks pale yellow. For many years this phenomenum created a great deal of concern amongst the local inhabitants but it was not until after the year 2000, as part of a nationwide campaign to examine the effects of mining, that environmental NGOs started using scientific data (already exisiting or derived from invited national and international scientists) to monitor the extent of the damage in San Sebastian and to publish the results. Some of these studies include: Barraza and Carballeira-Ocaña (n.d.) and Baker Matta et al. (200: 11, 14, 16) who analysed the presence of mercury, copper and arsenic in aquatic organisms in the Gulf of Fonseca (where the river drains its waters). Later Bianchini (2006) analyzed levels of manganese, iron, aluminium, copper, cyanide in river water as well as water Ph levels. Similarly Cartagena (n.d.) analyzed mercury and cadmium levels

---

[1]The existence of AMD in the mine was acknowledged by a company representative in an interview in the USA but he failed to admit any part in the process and instead attributed it to past mining operations. See the interview through this link: http://www.mptv.org/localshows/adelante/latest_episode/?v=8gRyq1wW8so&e=1311.

[2]From an administrative and legal perspective, what is clear from this episode is that government authorities at the time either failed to conduct an examination of the site or simply ignored the presence of AMD. In some of the RRA interviews the local inhabitants said that they started noticing a reddish tinge in the river water in the late 70s. According to some interviews the volume of water flowing in the river was much higher in the 50s so AMD discharge, if it was present, would not have been noticed without performing a dedicated test.

in the water in surrounding districts (also mentioned in Monroy 2008); Henríquez (n.d.) looked at manganese, iron, aluminum and copper levels in the soil; CEICOM (2010a) carried out a large study looking at aluminium, arsenic, lead, copper, iron, zinc, manganese, mercury, total solids, sulfates, borium, nickel and lithium in surface and ground water while Cortes (2010) studied levels of aluminium, zinc, lead, arsenic and cadmium in agricultural soils and maize seeds. The following year Larios et al. (2011: 9–10) studied arsenic levels in surface and ground water while MINEC (2011: 46–7) measured, water acidity, deforestation and rock slides. The government also carried its own study (MARN 2012: 5, 2013: 6) in order to measure levels of lead, arsenic and selenium in the soil and cyanide, arsenic, cadmium, aluminium, iron and water acidity in the San Sebastian River.[3]

The results from all these studies invariably tell the following story: the acidity levels in ground and surface water are high enough to not allow normal biological processes to occur and the levels of heavy metals sometimes surpasses nationally and internationally designated permissible levels. The latest government report has reiterated these results in a 2013 study. In it, MARN concludes that:

> The San Sebastian River was severely damaged by pollution from acid drainage emanating from the San Sebastian hill [where the mine is located]. In the dry season [the drainage] carries high levels of arsenic and cadmium and in the rainy season, high levels of cyanide, iron, aluminum and cadmium. All of these elements are toxic and precursor substances to environmental and human health problems due to their accumulation in organisms of aquatic life, and persistence in the food chain.

From the foregoing it is clear that the body of scientific evidence documenting the environmental damage caused by AMD in San Sebastian is robust and, equally important, that the government has accepted this evidence, carried out its own studies and has declared that San Sebastian requires further attention and some form of environmental remediation (PDDH 2016).

4.2 Socio-economic Legacies of the San Sebastian Mine

Even though environmental legacies in San Sebastian have been well documented, the same cannot be said about socio-economic legacies. Erzinger et al. (2008), ADES (2008), Larios (2008), Power (2008), CEICOM (2010b) and Mckinley (2012) have written studies and reports that go beyond environmental impacts and include both social and economic legacies at the national level. However, none

---

[3]Unfortunately during the course of the current study it was not possible to find any publication or report showing the contribution of heavy metals derived from pesticides used in the district.

provide enough relevant and detailed data about San Sebastian.[4] Given this situation, and in order to assess the socio-economic legacies of the mine, two questions had to be answered:

- AMD adversely affects ground and surface water but how does it affect people living in the surrounding community?
- How has the community been economically impacted since CGC abandoned the mine?

Let's start with the first question. The interviews during the RRA and the information from the secondary data analysis made it clear that the San Sebastian River was main source of water for the surrounding community and from the time its members became aware of the contamination they started finding alternatives to river water, and later, well water. Invariably these alternatives are costly from a household economy perspective and risky from a human health perspective. To be able to determine the impact of not having access to this water it was necessary to carry out an estimation of the market price of water in relation to household income in San Sebastian.

4.2.1 Estimating the Market Price of Water in Relation to Household Income in San Sebastian

For this task the study needed 3 basic pieces of data: household income, household water consumption and the market price of water per litre. In the absence of a dedicated income survey for San Sebastian and no district level desegregation of data in the national household survey the current study relied on data provided by the inhabitants of San Sebastian. From these interviews two hypothetical 5-member households were established: one with an income equal to that of the poverty line (US$1.25 per day and very close to the lowest reported income in the interviews) and one with an income of US$5 (a value between the top daily income reported in the interviews ($6 per day) and the minimum wage established by the government (US$4 per day).

From the interviews and the number of water tanks in the area, and in order to account for variability in household consumption, two different levels of

---

[4]A number of documentaries, short films and news items have been produced in Spanish dealing with socio-economic legacies. A selection of the most notable include:
"Oro sin Brillo" (https://www.youtube.com/watch?v=i8SiHX0pXOY), "Mineria San Sebastian" (https://www.youtube.com/watch?v=7kwnJFhSV3o),"Reportaje de mina San Sebastián en TV Izcanal" (https://www.youtube.com/watch?v=JLx8rYS5q8M), "Mina de San Sebastián en La Unión después de 100 años continua contaminando" (https://www.youtube.com/watch?v=BHN_srm6mpg), "El oro de Santa Rosa" (https://www.youtube.com/watch?v=2l-aOXU9Dvo) and "Las minas de oro de Santa Rosa de Lima" (https://www.youtube.com/watch?v=Ki7KQYKFRls).

consumption were established.[5] A "high" level consumption rural household based on the recommended World Health Organization (nd) daily levels (i.e.: human consumption of water without adverse long term health effects). The following list shows the estimates:

San Sebastian High Level Consumption Household (HLCH).

A. Drinking water, 3 lt (7.5% of total)
B. Cooking 5 lt (12.5% of total)
C. Bathing 15 lt (37.5% of total)
D. Washing utensils and clothes 7 lt (17.5%)
E. Toilet use 10 lt (25% of total)
Total: 40 lt per person per day.

The other household is a "low" level consumption rural household. The consumption levels for this household are based on some of the interviews during the RRA and national household consumption data. The list below shows the estimates:

San Sebastian Low Level Consumption Household (LLCH).

A. Drinking water, 1 lt (5% of total)
B. Cooking 1 lt (5% of total)
C. Bathing 10 lt (50% of total)
D. Washing utensils and clothes 3 lt (15%)
E. Toilet use 5 lt (25% of total)
Total: 20 lt per person per day.

The consumption estimates apply only to human consumption and do not include water for gardens, orchards, fruit trees and domestic animals because the variability of these factors would have skewed household comparisons. For instance some households have gardens and domestic animals whereas others do not. Gardens vary in size and the crop mix planted therein requires different amounts of water in order to thrive. The number of domestic animals (mostly pigs and chickens) varies greatly amongst households. From observations during the RRA, it is clear that both street vendors and stores sell commercially bottled water to people in San Sebastian. However, the calculations did not include its consumption as the

---

[5]According to the CARITAS PRA carried out at La Presa hamlet in 2014, there are 100 water tanks in the hamlet, some of them shared by more than one household. According to the same document out of 143 households, 100 have latrines and the rest have flush toilets. The current study was not able to identify if these and other sanitation facilities were "improved" or "unimproved" as defined by the WHO/UNICEF's Joint Monitoring Program for Water Supply and Sanitation. This would have had an effect on water usage calculations, as improved facilities tend to use more water to function. Even though El Salvador has experienced improvements in the area of access to sanitation facilities in the last 25 years, the gap between urban and rural areas is still wide. According to the WHO/UNICEF Joint Monitoring Program (2014), in 2012, 80% of the urban and 53% of the rural population had access to improved sanitation facilities. In the same year, 11% of the urban and 32% of the rural population had access to unimproved facilities.

interviews revealed that the amount of commercially bottled water households consume is negligible.

The water consumed for activities A, B, C, D and F above does not all come from the same source and does not cost the same. This required calculating prices for activities A and B separately from the others because water for the former needs to be potable and is therefore most likely to be purchased than water for the latter which, in most instances, is obtained at "no cost"[6].

In order to estimate the market price of water per litre this study relied on interviews with 2 community leaders, 3 artisanal miners and responses from the only 3 water vendors located in San Sebastian. The vendor's water prices per volume were verified with 2 wholesale suppliers in the nearby district of Limon, where San Sebastian water vendors buy their water at wholesale prices. Because the water volumes were reported in a mixture of Metric, Imperial and Spanish measures it was necessary to standardize all measures to the Metric system. The prices are therefore calculated on a per litre basis. The results from these calculations are as follow: the market price of potable water is US$0.015 p/lt. and the average price for non-potable water is approximately US$0.0022 p/lt.

4.2.2   The Cost of Obtaining Non-polluted Water in San Sebastian

As mentioned earlier, the study established two hypothetical household depending on their levels of consumption: a high level consumption household (HLCH, average: 40 lts of water per person/day) and a low level consumption household (LLCH, average: 20 lts of water per person/day). The findings show that for the former the cost is US$4.10 per m^3 with a total average monthly consumption of 6 m^3 while the latter pays US$2.50 per m^3 with a total average monthly consumption of 3 m^3.

In order to understand how expensive water is for these families the findings must be put in perspective: in San Salvador a 5-member household would pay no

---

[6]In the calculations carried out for the current study, "no cost" means that water from sources such as rivers, creeks and wells was not ascribed a monetary value despite the fact that the process of extraction and transport from these sources to the individual households represent no small amount of physical, time and energy expenditure. A similar approach was adopted with the cost of rain water stored in tanks because the costs of tank maintenance were difficult to quantify and integrate into the calculations. However, the consumption of this type of water does affect the annual household expenditure on water. This is specially the case during the rainy season when rain water stored in tanks, decreases the amount of water that needs to be purchased for activities A and B. Therefore the household annual water consumption calculations took into account the seasonal variation that exists between the rainy and dry seasons. For a whole discussion on the value of "no cost" water see: Cairncross and Valdmanis (2006). It is also important to note that the provision of water to households in a gendered activity. A study carried out in 18 municipalities in El Salvador (Halsband 1994) shows that women and children spent a great deal of their time supplying most of the water consumed in rural households.

Table 4.1 Water cost and consumption comparison in two districts of El Salvador

Location of 5-member household	Cost of water per m^3	Total average monthly consumption (m^3)
HLCH in San Sebastian	USD $4.10	6
LLCH in San Sebastian	USD $2.50	3
Average level consumption household in San Salvador	USD $0.025	27

Source Own estimates and Ibarra et al. (2005)

Table 4.2 Water cost and consumption comparison in two districts of El Salvador

Location of 5-member household	Average daily income	Percentage of income spent on water (%)
HLCH in San Sebastian	US$5.00	16.4
LLCH in San Sebastian	US$1.25	20.0
Average level consumption household in San Salvador	US$10.00	2.25

Source Own estimates and Ibarra et al. (2005)

more than US$0.25 per m^3. Such low unit price allows households in the capital city a total average monthly consumption of 27 m^3 (Table 4.1).[7]

The high cost of obtaining non-polluted water is most evident when the cost of water per m^3 is seen as a percentage of household income. HLCHs and LLCHs living on incomes of US$5.00 and US$1.25 per day use 16.4 and 20% of their respective incomes to buying non-polluted water in San Sebastian. In comparison, a San Salvador household living on an average daily salary of US$10 uses 2.25% of its income to satisfy its water needs (Table 4.2).[8]

How different is the situation in San Sebastian when compared to other places in rural El Salvador? Gasparini and Tornarolli (2006), found that that the poorest 20% of households in El Salvador (mostly located in rural areas) allocate around 10% of their income to satisfying water needs. With these statistics in mind, one can expect

---

[7]This figure is an average. Within San Salvador high volume consumption households use up a disproportionate amount of water in relation to other households. According to Ibarra et al. (2005: 32) "residential connections with a monthly consumption of 20 m^3 (ie 20% of connections and 7% of water consumed) have a daily per capita consumption of 81 L per day, while connections with a monthly consumption of more than 40 m^3 (21% of connections and 49% of water consumed) have a per capita consumption of 531 L per day. This last level of consumption is well above the average in OECD countries such as the Netherlands, Denmark and France (200 L per person) and Belgium, Germany and Portugal (120 L)."

[8]This calculation is based on a monthly charge of US$6.75 for 27 m^3 (or US$0.25/m^3). This is the m^3 rate charged by the state water supplier (ANDA) to an average household in San Salvador. However, private water companies can charge a similar household anywhere in the range of US $10.80 to US$37.80 a month. This means that households whose water service has been privatised pay between 155 and 544% more than if the water was provided by ANDA (Ibarra et al. 2005: 32).

that the majority of people living in rural areas would allocate a larger proportion of their income to pay for water than their urban counterparts and that those in San Sebastian are not allocating anymore than everyone else in rural El Salvador.[9]

However, the results from this study tell a different story. Even though the current RRA did not compare other rural districts with San Sebastian using the household water expenditure as a proportion of household income analysis, the figures from the current study show that people in San Sebastian allocated between 6 and 10% more of their income than their rural counterparts to buy water. The reason is that they cannot rely on the "no cost" water that is available to other rural communities in the form of rivers and wells. In other words, the community in San Sebastian pays an "AMD pollution premium" on top of the costs associated with being located in a rural area with no access to piped water.

Avenues for circumventing the pollution premium are few. Even though around 75–65% of the water consumed by San Sebastian households is obtained at "no cost", the price they have to pay for potable water is steep. To buy water for drinking and cooking alone, HLCHs and LLCHs devote around 12 and 10% of their respective incomes.

4.2.3 Health Problems in San Sebastian

Apart from building water tanks to catch rain water, people in San Sebastian also try to avoid paying the pollution premium by engaging in "risky" activities. These activities include washing and bathing in the river during the rainy season (when water volumes are high and the colour of the river looks almost normal), using river water to irrigate crops and gardens and allowing domestic pigs and chickens consume river water all year round. The health impacts of these activities are not perceptible in the short term but the long term risks are high. Studies carried out by CEICOM (2012: 6–7) and Nolasco (2012: 4) conclude that the mineral concentration in the river is many times higher than the permissible levels set by the PanAmerican Health Organization and the World Health Organization. Such high concentration of minerals "can cause different types of cancer in people exposed to water use and consumption, as well as gastrointestinal, cardiovascular and nervous system dysfunction" (MARN 2013: 6). For instance, CEICOM (2010a) and Cortes and Díaz (2011) have carried out separate studies in which they argue that the presence of heavy metals in San Sebastian may be contributing to the high incidence of Guillan-Barre Syndrome in the area. According to the WHO,

---

[9]Other statistics confirm that people in rural areas and the poor are less likely to have access to clean water. According to DIGESTYC (2014) only 25% of rural households have access to potable water through a pipe (which is more likely to be less expensive per volume) and Moreno found that 68% of those categorized as living in extreme or relative poverty do not have access to piped water (2005: 13). It is not surprising that the UNDP ranks El Salvador in third place amongst Latin American countries with the highest level of inequality of access to drinking water (2010: 39).

Guillan-Barre Syndrome has an annual incidence of 1–3 people per 100,000. In 2011, the government health clinic in San Sebastian reported 5 cases of this syndrome which is relatively high given that the population of the district at the time was around 3000 and the population of the Department of La Union, where San Sebastian is located, did not surpass then 230,000 inhabitants (Cortes and Díaz 2011: 101; MINSAL 2015). A look at the table below shows that the incidence of Guillan-Barre is high in the Eastern departments of El Salvador where mining and high use of pesticides has historically taken place (Table 4.3).

Another potential pathway affecting humans is the food chain. After examining AMD data from Larios (2008) and Bianchini (2006) Noller from the Centre for Mined Land Rehabilitation[10] said that "the metals will all be soluble and be taken up easily by plants and other aquatic and terrestrial species. There is a high risk of bio-accumulation via the food chain".[11] It is also important to note that the use of pesticides in the district represents an additional, albeit smaller, heavy metal burden to the food chain via the river and soil but the current study could not find any analysis that disaggregated the sources of heavy metals in the district.

Apart from being in direct contact with AMD and exposed to the toxicity of heavy metals there is another type of risk that people in San Sebastian face. From the RRA interviews it became clear that some people do not drinking enough water according to WHO standards. Low water intake can lead to adverse long term health effects and/or accelerate the progression of already existing health problems such as Chronic Kidney Disease (CKD) which is already prevalent throughout the eastern part of the country where San Sebastian is located. As Hernadez says:

> The factors of progression of CKD are [among others]: excessive protein intake, obesity dyslipidemia, obstruction and urinary tract infections, smoking, long hours under sunlight exposure, *poor water intake*, medication and exposure to *nephrotoxic products including pesticides* (Hernández et al. 2011: 7–8, my emphasis).

However, this type of risk taking is not unique to San Sebastian. According to Cairncross and Valdmanis (2006) an individual's daily water consumption in rural areas depends largely on the proximity of water sources. When a person needs more than 5 min to obtain water then the tendency is to consume only what is necessary to survive. This type of risk is assumed by the local population in order to save money, time and energy in the short term and could be avoided if the Salvadorean health authorities provided accessible water for the community.

In contrast, avoiding the legacies of AMD contamination requires toxicology studies to be undertaken so that toxicity pathways showing how heavy metals found in the San Sebastian River travel through land, air and water into animals and plants and ultimately into humans are understood. The great limitation for health authorities in El Salvador is the complete absence of this type of studies which

---

[10]Personal communication via e-mail, September 8, 2014.

[11]The data in these studies examine the large increase in sulfate, decreasing pH and increasing all dissolved metals located down river from where the mine is and close to the mine's water discharge point.

Table 4.3 Registered cases of Guillan-Barre Sindrome according to Department in Eastern El Salvador (2003–2014)

Departments	2003	2004	2005	2006	2007	2008	2009	2010	2011	2012	2013	2014	Total
La Union	19	12	9	3	6	4	3	8	12	8	4	12	100
Morazan	7	2	2	5	3	4	4	4	5	5	4	4	49
San Miguel	26	13	19	8	4	11	10	19	14	15	11	10	160
Usulutan	7	7	10	2	3	12	8	7	7	9	12	12	96
Region Total	59	34	40	18	16	31	25	38	38	37	31	38	405

Source SIMMOW database, Ministry of Health (2015)

means that until that situation is remedied they will only be able to treat the symptoms of desease presented by patients and not its etiology.[12]

4.3 Lost Opportunities

When examining how the community has been economically impacted since CGC abandoned the mine it is clear from the previous section that AMD contamination costs a significant amount of money to individual households in San Sebastian. However, there is a distinction between legacies where income is foregone in order for households to continue functioning as normal (as is the case of having to pay to obtain non-polluted water) and other legacies where communities (and countries) suffer opportunity losses and/or uncertainties as a result of the cessation of mining activities. In order to understand the latter type of legacies, these opportunities must be explained.

4.3.1 Economic Opportunities Associated with Mining and Other Natural Resource Extraction Activities and the Impact of Abandonment

Despite the "resource curse" literature,[13] the ICMM argues that mining and other natural resource extraction industries can offer the following contributions to the economies of the countries in which they operate:

- Foreign Direct Investment (FDI). This is mostly made up of the initial investment made by foreign companies setting up extractive projects. From a total amount of contributions by a single project, FDI's share ranges from 60 to 90% of contributions to the host country's macroeconomy.
- Exports. Depending on the volume and value of exported commodities, the host country's macroeconomy benefits from increased export revenues and improved balance of payments situation. From a total amount of contributions by a single project, export's share ranges from 30% to 60% of contributions to the host country's macroeconomy.

---

[12]The head doctor at the local government clinic confirmed this when he said that they treat patients symptomatically and that without the required studies it is nearly impossible to establish the cause of the symptoms presented.

[13]There is substantial evidence in the natural resource literature to suggest that the economic contributions made by extractive industries to a host country are negated by what is commonly referred to as "the resource curse". In relation to the impact of natural resource endowments on economic outcomes see: Rosser (2006). However, in relation to their impact on social outcomes, Edwards (2016: 1) argues that the "evidence [...] remains thin" but points to Van der Ploeg (2011) for a review of the literature.

- Government revenues. Depending on the level of effective taxation and royalty rate, the host economy benefits from increased inflows into the state's treasury. From a total amount of contributions by a single project, taxation's share ranges from 3 to 20% of contributions to the host country's macroeconomy.
- National income. Depending on the complexity and duration, the capital costs involved in running an extractive project, (such as those devoted to procurement and infrastructure), can be significant contributors to the national as well as the local economy. From a total amount of contributions by a single project, capital cost's share ranges from 3% to 10% of contributions to the host country's microeconomy.
- Employment and community expenditure. Since most formal contemporary extractive projects are capital intensive, employment tends to be low and highly skilled which may, if the necessary conditions exist, provide opportunities for local participation in a project. If there is no government legislation determining the amount and type of community expenditure, then what the host community receives will depend on its bargaining power vis-à-vis the project operator. In any case the amount provided for community expenditure is invariably lower than amount provided in the form of wages. From a total amount of contributions by a single project, employment and community expenditure's share ranges from 1 to 2% of contributions to the host country's microeconomy (ICMM 2011).[14]

While CGC operated in El Salvador some of the above mentioned benefits were realized but their quantification is difficult because CGC figures on payments to government, employment and exports are not available so government statistics are used instead.

Since the 1960s, El Salvador's mineral production represents only a fraction of the national economy and the general long term trend is one of decline. In 1995, a typical year for the industry, production "accounted for less than 1% of [El Salvador's] estimated gross domestic product (GDP) of US$10 billion" (Doan 1995: 1). According to El Salvador's Central Bank, the mining industry suffered an average decline of 5% in production output during the period 1990–2013. However, within the general declining trend, the industry is characterized by highs and lows. For instance, in 1990 total production was worth US$18.1 million, increasing to US $36.2 million in 2003 before declining to US$22.8 million in 2010. In the latest reported year, 2014, production reached US$69.9 million, or 0.3% of GDP (PDDH 2016).

The contribution of gold exports to mineral exports in the last 3 decades has been minor, averaging 1% of total mineral exports. In 2005, for instance, when the

---

[14]To this list, the IFC includes compensation payments as contributions to the local economy but one cannot see how this item can be regarded as such since this type of payment is made to compensate for the loss of something that the community already possessed, usually in the form of natural assets, land and housing, before extraction occurred. Therefore, compensation should not be classified as a benefit from resource extraction.

total value of mineral exports reached US$32 million, the value of gold exports equaled US$0.3 million (PDDH 2016). Based on average production figures for the late 1990s CGC would have exported around US$300,000 per annum until 2000, when the company stopped gold production.[15]

Based on figures provided by CGC, the company claims to have made FDI contributions totaling US$88.3 million for the period 1987–2004 or approximately US$5.2 million per annum in the 17 year period (SEC 2004). However, CGC seems to have stopped making major investments in El Salvador as early as 2000 when it stopped producing gold. FDI statistics confirm this observation. In 2004, total FDI in the mining sector did not reach the US$ $1 million mark compared to US$363 million going to the other economic sectors (Andersen 2004: 6). The amount of FDI going to gold mining would have been significantly less than US$1 million since in that year there were other gold mining companies apart from CGC investing in exploration activities. Also that figure includes FDI for the entire minerals industry, which in El Salvador, is made up of companies involved in the extraction and production of gypsum, mineral fertilizers, lime, limestone, sand and gravel, hydrocarbons and some metallic manufacturing. One can deduce from the foregoing that after CGC stopped producing gold, its level of investment dropped to a miniscule amount in the context of the Salvadorean economy.

In terms of company payments to the government, there are records on rents but not on royalties paid.[16] Of note is the variability of the amount of rent paid. The figures from 1998 to 2002 move from US$600 at the beginning of that period to a high of US$1800 in 2000 to a low of US$150 at the end of the period. Until 2000 the corporate and royalty rates were set at 25 and 4% respectively (Velasco 2000: 6). Unfortunately, no profit figures were available from company records, without which it is not possible to calculate the corporate tax rate. Based on average production figures the royalty rate paid would have yielded US$12,000 per annum from 1987 to 2000.[17]

In terms of national income it was not possible to ascertaian the amount CGC spent on local procurement and infrastructure or if it had any subcontracting arrangements with local suppliers. The only reference found in this regard comes from an annual report informing that from 1987 to 2004 the Joint Venture and its subsidiaries spent US$89.2 million to support the operations at the mine. The report contains a long list of expenditure items but no breakdown of how much was spent

---

[15]Calculated at 3000 oz per annum at a cash cost of US$100 per ounce.

[16]Partly this is the result of the way Salvadorean freedom of information legislation is set up. An individual is entitled to find out about taxes paid at the Departmental level but the government will not provide information about taxes paid by individual companies. The information provided here is therefore based on taxes paid by the mining industry in the Departments of La Union and San Miguel, knowing that CGC was the only mining company operating in those departments.

[17]From 2000 the royalty rate was reduced to 2% and according to the current mining law, half of that must be directed to the municipality in which the mine is located.

on each.[18] This information would have been vital in desegregating the company's economic impact on the surrounding community from that occurring at the level of the national economy.[19]

The figures for employment provided by the company vary according to source and period. Ex-employees reported during the RRA interviews that, at its height in the 1990s, CGC employed at least 130 workers, most of who were engaged in manual labour. A 2004 report states that the company was employing 35–45 workers at the time of writing (SEC 2004). In terms of community investments, the same report declares that CGC spent money on "the construction of a community bridge, for a community telephone building and facilities, [and] a community place of worship" but the report did not reveal a breakdown of costs nor a total for these items.

In sum, from available figures, the overall economic contribution made by the company in the period 1987–2004 to the Salvadorean economy is approximately US$92.2. Of this total, no less than US$3.6 million (approximately US$211,000 per annum) would have been spent in and around San Sebastian. Once mining ceased in the district, and without any other economic activity capable of absorbing the unemployed workers from the mine, they turned, in increasing numbers, to artisanal mining as a substitute, and familiar, income generating activity. More in relation to this phenomenon will be examined at the end of this chapter.

Generally, the loss of opportunity type of socio-economic legacy is very difficult to avoid in mining projects because of the finite nature of resources, unless ongoing investment is made in new or existing economic sectors not linked with mining. It would take investments in other economic sectors equivalent to those made by CGC during the production phase to recover the opportunity losses in San Sebastian.[20] A different approach would have included educating and training local

---

[18]The reports says: The funds invested in the Joint Venture were used primarily for the exploration, exploitation, and development of the SSGM, for the construction of the Joint Venture laboratory facilities on real estate owned by the Company near the SSGM site, for the operation of the laboratory, for the purchase of a 200-ton per day used SCMP precious metals' cyanide leaching mill and plant, for the initial retrofitting, repair, modernization and expansion of its SCMP facilities, for consumable inventory, for working capital, for exploration and holding costs of the San Felipe-El Potosi Mine, the Modesto Mine, the Hormiguero Mine, and the Montemayor Mine, for SSGM infrastructure, including rewiring, repairing and installation of about two miles of the Company's electric power lines to provide electrical service, for the purchase of equipment, laboratory chemicals, and supplies, for parts and supply inventory, for the maintenance of the Company-owned dam and reservoir, for extensive road extension and preservation [...] for the purchase of the real estate on the Modesto Mine, for leasing the Montemayor real estate, for the purchase and erection of a cone crushing system, for diamond drilling at the SSGM, for the purchase of a rod mill and a carbon regeneration system, for holding costs, and all other related needs (SEC 2004).

[19]It must also be noted that this figure includes the FDI mentioned above and should not be double counted.

[20]It was beyond the scope of this study to examine the legacies of migration to or from this mining district. Unlike other mining districts in Latin America, San Sebastian seems not be experiencing a sudden influx of migrants from other parts of the country to work the mine and this may have to do

human resources with transferable skills at the start of the extraction project so that these become employed in existing or new sectors of the economy. These and other strategies will be further explored in the following chapter.

4.3.2   Uncertainty in Relation to Land Tenure

According to data provided by the National Registry Center (CNR), the land in which the mine is located has an area of approximately 2.1 km^2 and is owned by San Sebastián Gold Mines Inc, one of CGC's subsidiaries. This company also leases land from Minsanse which is partly CGC owned and partly worker owned. During the interview process many informants reported not to have any certainty to ownership of the land in which they were living despite the fact that some of them are part owners of Minsase.[21] Many of them argued that this provision only provides a temporary solution to these households because the land on which they have built their homes technically does not belong to them, and they would prefer to have more security by having legal title to the land. This situation is so problematic in El Salvador that it has been covered in the national press. El Diario Latino interviewed Mr. Gustavo Blanco, who has lived on the company's property since he was a child:

> This company was founded with 90 workers so there is only one social deed in common, but no one can get a registered title for land […] if the owners return and they find a vein of gold in my lot, I would not have nothing to do here because the company [CGC] is legally the owner (Diario CoLatino 2014).

Without title to the land, the inhabitants fear that they can be accused of squatting on CGC land and claim that they cannot access government services, such as potable water from the state water authorities.[22]

(Footnote 20 continued)

with their part ownership of Minsanse but the author could not verify this hypothesis. Like many other towns in El Salvador, many of San Sebastian's residents have opted to leave the district in search of better opportunities elsewhere. Most of the artisanal miners live either in San Sebastian or they travel from adjacent districts to work at the mine.

[21]The total number of shareholders is not clear and varies according to source. In the interviews some ex-workers declared that the Minsanse started with 90 workers, each owning 4 shares and the (then) owner of the mine keeping 90 shares. Other interviewees stated that there were 138 shareholders based on the fact that at one point in time the company employed that number of people in the mine. In 2004 CGC said that there were "around 200 unrelated shareholders" (SEC 2004). In all likelihood, if all workers in Minsanse were allocated shares that they could keep, the number of shareholders increased over time, even if the same number of workers were no longer employed at the company. What is clear, however, is that the workers are not majority shareholders.

[22]Despite enquiries at the CNR, it was not possible to clarify the current land tenure situation in San Sebastian. When the author sought an answer from the Salvadoran Institute of Agrarian Transformation (ISTA) officials there said that they could not help San Sebastian residents because the area of the property does not exceed the legal limit allowed for ISTA to initiate a process of expropriation of CGC property.

During the course of this study it was not possible to quantify this legacy but from the above it is clear that this type of situation results in economic insecurity for the residents living within the mining concession. Since there is no clarity as to whether the land on which they live is legally theirs, this represents another form of opportunity loss directly related to the way CGC abandoned the mine.

Apart from the opportunity losses described above there is an additional socioeconomic legacy that is present in San Sebastian and it has to do with the growth of artisanal mining upon CGC's departure.

4.4 The Growth of Artisanal and Small Scale Gold Mining (ASGM) in San Sebastian

Currently in El Salvador artisanal miners or *Güiriseros* as they are known locally, are only active in the San Sebastian region. Official data on the *güiriseros* exists but it is not extensive. Reports from MARN (2012, 2013) mention them in passing as does a 2011 MINEC strategic environmental evaluation on the mining sector (pages 25, 27 and 47). Another report from the same ministry (MINEC 2012) focuses on their numbers and activities but not on their gold output. What else is known about this group originates from studies carried out by NGOs and academic researchers who mention them briefly (Kingbury 2014: 12 and 26; CEICOM 2010a: 6; Larios 2008: 1,3, 9 and 10; Cortes and Díaz 2011: 74; López et al. 2011: 10). Doan (1995), a contributing author of the USGS Minerals Yearbook devotes 5 lines of text describing their activities. The Salvadorean media has also run brief stories about their situation.[23]

From the RRA interviews with artisanal miners, ex-CGC workers and data from Doan (1995), it is clear that ASGM has been taking place in San Sebastian for some time and includes the period when CGC was the main operator. The relationship between *güiriseros* and CGC has not been made clear in the literature and maybe relates to Minsanse being worker owned and holder of a mining license. This status allowed those Minsanse small shareholders who could not find formal employment with CGC to try their own luck in marginal sites of the mine and thus become "legal" *güiriseros*. While GCG operated the mine, the number of *güiriseros* was kept to a minimum but this changed after the company abandoned the mine. Their growth in the last few years is the direct result of unplanned mine closure and this has led to a number of environmental and socio-economic legacies. Before looking at San Sebastian's *güiriseros* in particular it is necessary to put them in context so that a more nuanced understanding of ASGM in El Salvador can be gained.

---

[23]For instance, see: Vallejo (n.d.).

4.4.1 ASGM and Its Status in El Salvador and Beyond

In economic terms, the extraction of gold can be categorized as either industrial large scale and formal or artisanal small scale and informal. The former enjoys state recognition, operates with large amounts of capital, modern technology and is carried out by private or state owned companies in most cases. In contrast, the latter usually does not enjoy state recognition, operates using traditional methods of extraction that requires little capital and is carried out by small groups of individuals, families and sole miners. Sometimes the income generated from mining is enough to support a family but many times this income is so small as to be regarded a subsistence activity that has to be complemented with other economic activities. The fact that artisanal mining is small in scale, informal and sometimes illegal at the production stage does not mean that its contribution to formal economic activity is insignificant. According to the UNIDO/UNDP/GEF Global Mercury Project:

> At least 100 million people in over 55 countries depend on ASGM – directly or indirectly – for their livelihood. ASGM is responsible for an estimated 20–30% of the world's gold production, or approximately 500–800 tonnes per annum. It directly involves an estimated 10–15 million miners, including 4.5 million women and 1 million children (Telmer 2008; UNEP 2008 cited in UNEP 2009: 7).

Like in many other parts of the world, artisanal miners in El Salvador are considered informal and the law renders their activities, since the 2008 mining ban, illegal. Despite their legal status, they have grown to become an integral part of the larger national and international socioeconomic landscape. San Sebastian's *güiriseros*, as producers, form part of extensive and complex global value chains that include suppliers (such as pharmacies, hardware stores, fuel stations, and metal fabricators), intermediaries (such as local gold buyers, pawnshops, jewelry shops and smelters) and exporters within national, regional and international markets. Unlike many other minerals, gold is highly portable (small amounts have high market value), fungible (one unit can be substituted with another of the same grade) and liquid (can be readily exchanged for money) close to the point of extraction or recovery. These features and the fact that the metal is relatively easy to process from ore to its pure form allow gold artisanal miners to keep a relatively modest portion of the international market price of each ounce of gold extracted in comparison to what artisanal miners keep when engaged in the extraction of low value minerals (Hayes 2013).[24] Interviewed San Sebastian *güiriseros* reported selling one

---

[24]One reason for this may be that the global gold value chain is producer dominant. According to Gereffi and Korzeniewicz (1994) there are two types of global value chains: buyer dominant and producer dominant. In the former buyers have enough power to set some market parameters while in the latter the power lies with the producer. Even though artisanal miners are not powerful in themselves, each miner forms part of a larger group of formal gold producers that include small, medium and large scale mining companies who are able to exert a great deal of influence on the final price of gold.

gram (0.03 oz) of gold for US$32 when "the price is low and USD $50 when the price is high".[25]

4.4.2 The Use of Mercury in San Sebastian

The *güiriseros* who took part in the RRA interviews expressed that mining is an inherently dangerous and risky activity and seem to think that the many accidents that occur at the mine (mostly as a result of the absence of safety procedures and proper equipment) are normal. Apart from accidents they also expressed concern about substances that may affect their bodies but showed absolutely no knowledge of the fact that mercury is a toxic substance. The *güiriseros* use basic tools to extract the ore, which is then crushed, transported to their homes and separated through a number of processes that include the use of mercury. The RRA interviews confirmed that the *güiriseros*:

1. Buy mercury either from a local pharmacy or the contraband market. The former source is preferable because it is less expensive than the latter;
2. Recover less and less mercury with each use (a sign of evaporation);
3. Use less mercury when they add other chemicals to the ore when processing and;
4. Use cotton fabric to separate the mercury from the gold;
5. Handle mercury without any form of protection and in the presence of others (including members of the family).

Unfortunately during the course of this study it was not possible to determine the exact amount of mercury used and its subsequent discharge into the environment. The author therefore decided to calculate usage based on the production of gold. Although the *guiriseros* were not willing to disclose the exact amount of gold they sometimes sell, they provided enough data to estimate minimum and maximum amounts of extracted gold. From their reports, it is estimated that the average amount of gold produced by each *güirisero* is 60 grams per year.[26] MINEC (2012) has estimated that there are about 400 guiriseros in the area, but this figure varies with the seasons and only a fraction of this number works full time in the mine at any one time. For most, income from mining is a complement to other income-generating

---

[25]The international price of gold at the time of interviews was approximately US$55 per gram.

[26]According to Domínguez et al. the *güiriseros* extract some 40 kg of gold annually from the mine (2015: 170). But whereas these authors use a monthly salary of USD$ 300 to arrive at that figure, the current study used an average salary of USD$ 150 and also calculated the data according to the minimum amounts (1 g) and maximum (120 g) of gold extracted per year reported by those taking part in the interviews. This figure was multiplied by the total number of *güiriseros* in the area. The minimum total amount per year is 4.8 kg while the maximum is 48 kg. The average of the two figures is 24 kg.

activities. However, with the reported figures it can be said that the combined amount of gold mined in the area is 24 kg per year. This means that to process their annual output the *güiriseros* use approximately 720 kg of mercury which, through evaporation and use, is dispersed over time in the environment.[27] In comparison, a UNDP assessment for Central America and the Caribbean estimates that artisanal miners consumed between 15 and 30 metric tonnes of mercury in 2005 (UNDP 2009: 13) making ASGMs one of the largest users of mercury in the region.[28]

It is worth noting that not all the mercury used in the extraction and cleaning process is dispersed in the same year. Veiga and Baker have estimated that when no retorts are used in the extraction process (which is the norm in San Sebastián), the proportion of mercury dispersed in the environment can reach up to 45% of the mercury used and the rest is reused or recycled until eventually everything is lost (2004: 19). The same authors report that in the region of Callao, Venezuela, the proportion of mercury lost in gold production can be 1.5–3, that is, for each gram of gold produced, 1.5–3 g of mercury are dispersed. In the absence of a constant proportion between the amount of gold produced and the amount of mercury dispersed in the environment it can be speculated that the amount of mercury that is dispersed every year in San Sebastián can vary from 36 to 72 kg if using the proportions 1:1.5 and 1:3 respectively.

The above figures are relative to who and how mercury is handled. The people most at risk of inhaling mercury vapors are the *güiriseros* and their families, jewelers and any person living nearby where gold is melted with traces of mercury because the concentrations of mercury are high. Veiga and Baker cite a gold-melting studio where an average concentration of 83 mg Hg/m^3 was recorded

---

[27]To calculate this figure it was assumed that no more than 30 grams of mercury is needed to process one gram of gold. This assumption is based on the fact that the mercury smuggling price is USD$ 5 per ounce (28.3 g) which means that this input represents 15% of the USD$ 32, which is the price that the guiriseros received for a gram of gold during the low price period. If the *guiriseros* were to pay more for the mercury, the business of gold extraction would not be economically viable. Therefore, they may be using less mercury than calculated but the exact amount can only be measured by conducting a specialized study that determines the type of process they use and the concentration of gold in the ore they work on. The reason for this is that there is no constant ratio between the amount of mercury used and the amount of gold extracted. Veiga and Baker, for example, cite studies of sites where miners can use just one gram of mercury to draw one gram of gold (2004: 18). In other places more is needed. In comparison, official statistics indicate that El Salvador imported 60 kg of mercury in 2014 and 1246 kg in 2008 to be used in the manufacture of chlorine-caustic soda compact and traditional fluorescent lights, dental amalgam, neon lights, button-type batteries, cleansers, fireworks, folk medicines, grandfather clocks, pesticides, medical devices, thermostats, car switches and some skin-lightening creams and soaps (USGS 2012). However, only a fraction of the official imports ends up in San Sebastián.

[28]In the same region and year, dental applications, measuring and control devices and chlor-alkali production consumed between 20–25 tonnes, 10–15 tonnes and 5–15 tonnes respectively. In El Salvador, official statistics indicate that, in 2008, the country imported 1246 kg of mercury to be used in the manufacture of chlorine-caustic soda compact and traditional fluorescent lights, dental amalgam, neon lights, button-type batteries, cleansers, fireworks, pesticides, medical devices, thermostats, car switches and some skin-lightening creams and soaps (USGS 2012). However, only a fraction of the official imports ends up in San Sebastián.

for a period of two hours after melting gold with traces of mercury. The same authors caution that in many of these places "the ventilation systems used to extract the vapors are generally rudimentary and consist only of a fan [which usually] blows Hg vapors into the urban atmosphere, exposing people living nearby, which is considered extremely dangerous" (2004: 18).

4.4.3 Mercury and Its Problems

This heavy metal may be associated with the environmental impacts in eastern El Salvador described by Baker Matta et al. (2002), Barraza and Carballeira-Ocaña (n. d.) and human health problems described by Larios (2008), Cortes (2010), Cortes and Díaz (2011), CEICOM (2010a, b) and Cartagena (n.d.) but so far no study has shown that the presence of mercury in San Sebastian is the cause of human illnesses and no case of mercurialism has been recorded so far by the health authorities.[29] To do that, more studies need to be carried out since mercury's impacts on humans are particularly difficult to trace. Mercury, in its elemental form, is poorly absorbed and poses low risks to humans. Mercury vapor, on the other hand, is more easily absorbed and is harmful at high levels. Mercury in its methylated form is highly toxic, but its transfer to humans depends on many factors. As Veiga argues:

> The mechanisms and rates of accumulation and elimination are unclear, but appear to depend on the specific biological characteristics of each species of fish as well as the properties of the aquatic systems. A comparison of animals differing in species, size and feeding habits confirms that the food intake of Hg is far more important than direct uptake from water (Veiga 1997: 9).

Once mercury in its vapor or methylated form has been absorbed by the body, its harmful effects can be difficult to diagnose. In this regard, Veiga argues: "a person suffering from a mild case of Hg poisoning can be unaware because the symptoms are psychopathological. These ambiguous symptoms may result in an incorrect diagnosis" (1997: 6–7). Therefore, despise the fact that the Ministry of Health, reported 12, 293 first time consultations as a result of acute infections of the upper respiratory tract at the San Rosa de Lima clinic in 2013 (MINSAL 2015)[30] and that medical evidence shows that "widespread exposure to mercury causes intense irritation in the respiratory tract, producing bronchitis, pneumonia, and bronchiolitis" Cortes and Díaz (2011: 75) warn that the causes of the infections experienced by San Sebastian residents are diverse and can range from "a sudden change of temperature to heavy metal contamination" (2011: 75). In addition, the RRA interviews reveal

---

[29]Cartagena (n.d.) maintains that heavy metals contamination may be contributing to the high prevalence of CKD in eastern El Salvador where mining has taken place for decades.

[30]The ministry's database (SIMMOW) uses the international diagnosis code J00-J06-9 to classify these illnesses. In the same period the next most common reason for visiting the doctor at the clinic was urinary tract infections, with 1976 cases reported.

that health and environmental officials stationed in Santa Rosa de Lima do not have the capacity to deal with the diagnostic complexities that Veiga is referring to.

The difficulties in diagnosing mercury poisoning in individuals and communities are compounded by two factors related to this heavy metal. The first has to do with bioaccumulation's time/space characteristics. The process of accumulation of heavy metals in living tissues takes a long time and therefore the ill effects may show up years later and sometimes in the next generation. Veiga, for instance, found that in Japan, "workers with a peak urinary Hg concentration of 600 µg/l showed neurobehavioral disturbances 20–35 years after the mercury vapor exposure" (1997: 6–7). Mercury found in the food chain may have bioaccumulated via species that can cover large geographical areas and the final consumption may take place in a location very distant from the original mercury source or handling point. Baker Matta et al. (2002), for example, argues that the source of heavy metals found in the shellfish at the Gulf of Fonseca[31] may be originating from the various mining districts located 20–100 kms away. The second factor has to do with selenium and its antagonistic effect against mercury assimilation in living tissue. Studies by Belzile (2014: 129) have shown that there is an inverse relationship between concentrations of mercury/methyl mercury and selenium in aquatic organisms. Selenium is found in San Sebastian soil and may be contributing to reducing mercury assimilation in local species, but no study has been carried out in the district to confirm this hypothesis.

From the foregoing it is clear that since no toxicity pathway study and serology of the region (to ascertain the transport of mercury and its level in human blood) have taken place in San Sebastian there is no way of knowing the health effects mercury may be having on the population. However, this situation does not rule out that the handling of mercury may be affecting the long term health of the *Güiriseros*, their families and the environment.

It is also important to note that despite the problems associated with the *Güiriseros*' activities it is necessary to understand that, in the long term, a simple ban on their way of life is unlikely to work for the state and the community. Many authors (Davidson 1993; Veiga 1997; Hentschel et al. 2002; Chaparro Avila 2001) have suggested that ASGM can either be formalized or transformed so that artisanal miners can transition to a sustainable livelihood as part of a long process of mine closure which is a topic that will be examined in more detail in the following chapter.

References

ADES (2008) Minería metálica y su inviabilidad en El Salvador. Santa Marta. Documento de discusión. Asociacion de Desarrollo Economico y Social

---

[31]The Gulf of Fonseca is where the San Sebastian River drains its waters.

Andersen S (2004) The mineral industries of Central America: Belize, Costa Rica, El Salvador, Guatemala, Honduras, Nicaragua and Panama. Washington, USGS. Retrieved December 21, 2016 from https://minerals.usgs.gov/minerals/pubs/country/2004/camermyb04.pdf

Baker Matta M, McKinnie D, Barraza E, Sericano J (2002) Hurricane Mitch Reconstruction/Gulf of Fonseca Contaminant Survey and Assessment. Maryland, National Oceanic & Atmospheric Administration

Barraza J, Carballeira-Ocaña A (n.d.) A short survey of heavy metals from La Union Bay, Gulf of Fonseca, El Salvador. San Salvador, Ministerio del Medio Ambiente y Recursos Naturales

Belzile N (2014) Minerals, Metals, Toxicity and Substitutes. In: Leadbeater D (ed) Resources, empire and labour: crisis, lessons & alternatives. Fernwood Publishing, Nova Scotia

Bianchini F (2006) Estudio técnico. Calidad de agua en la zona de explotación minera de San Sebastián, Municipio de Santa Rosa de Lima, Departamento de La Unión. San Salvador, CEICOM

Cairncross S, Valdmanis V (2006) Water Supply, Sanitation, and Hygiene Promotion. In: Jamison DT, Breman JG, Measham AR (eds) Disease Control Priorities in Developing Countries, 2nd edn. Washington (DC), The International Bank for Reconstruction and Development/The World Bank; New York, Oxford University Press

Cartagena R (n.d.) Volcanes, Mineria de oro e Insuficiencia Renal En El Salvador. Unpublished manuscript, San Salvador

CEICOM (2010a) Análisis de la calidad de agua y su relación con la salud y calidad de vida de los pobladores del Río San Sebastián, en la zona de minas San Sebastián. San Salvador, CEICOM

CEICOM (2010b) La minería y sus impactos sobre la agricultura, los recursos hídricos y la salud humana. San Salvador, CEICOM

CEICOM (2012) Contaminación e Impactos de la Minería Metálica en el Cantón San Sebastián. San Salvador, CEICOM

Chaparro Avila E (2001) Taller Preparatorio de la Conferencia Anual de Ministros de Minería/Memorias Caracas. Serie Seminarios y Conferencias, 16. Santiago de Chile, CEPAL

Cortes C (2010) Determinación de metales pesados en suelos, sedimento y semilla de maíz, en áreas agrícolas expuestas a desechos de minería metálica, y su repercusión en la cadena alimenticia. San Salvador, CEICOM

Cortes C, Díaz CI (2011) Caracterización de la mortalidad vinculada a impactos ambientales en los pobladores que viven alrededor de la mina San Sebastián, Santa Rosa De Lima, Departamento de La Unión. Tesis de Posgrado, Escuela de Ciencias Sociales, Maestría en Métodos y Técnicas de Investigación Social, San Salvador, Universidad de El Salvador

Davidson J (1993) The transformation and successful development of small-scale mining enterprises in developing countries. Nat Resour Forum 17(4):315–326

Diario CoLatino (2014) La minería metálica, el conflicto socio – ambiental por el agua. 22 de Abril

DIGESTYC (2014) Encuesta de Hogares y Propositos Multiples, 2009–2013. Direccion General de Estadisticas y Censos, San Salvador

Doan D (1995) The mineral industry of El Salvador. Washington. USGS Minerals Information. Retrieved December 21, 2016 from https://minerals.usgs.gov/minerals/pubs/country/1995/9539095.pdf

Domínguez JP, Moya M, Rodríguez E, Panameño P, Linares J (2015) Evaluación final de riesgos y propuesta de medidas de remediación en 15 pasivos ambientales mineros de El Salvador. Fundación Maquilishuatl, FIAES, MINEC, San Salvador

Edwards R (2016) Mining away the Preston curve. World Dev 78:22–36. doi:10.1016/j.worlddev.2015.10.013

Erzinger F, Gonzalez L, Ibarra AM (2008) El Lado oscuro del oro. Impactos de la mineria metalica en El Salvador. San Salvador, CARITAS de El Salvador and Unidad Ecologica Salvadoreña

Evanhoe R (2006) Metal pollution from medieval mining persists. Chemical and Engineering News. August 17

Gasparini L, Tornarolli L (2006) Disparities in Water Pricing in Latin America and the Caribbean, Human Development Occasional Papers (1992–2007) HDOCPA-2006-22, Human Development Report Office (HDRO). United Nations Development Programme (UNDP), New York

Gereffi G, Korzeniewicz M (eds) (1994) Commodity chains and global capitalism. Greenwood Press, Connecticut

Halsband S (1994) Diagnóstico sobre la participación de la mujer en la obtención del agua. El Salvador, Organizacion Panamericana de la Salud

Hayes K (2013) Legal and market-side demands for traceability in the mineral supply chain. Paper presented at the International Conference on Artisanal and Small-scale Mining in the Asia Pacific Region: Current Status and Challenges. Ulaanbaatar, Mongolia, 28–31 May 2013

Hentschel T, Hruschka F, Priester M (2002) Global report on artisanal and small-scale mining. London, Mining, Minerals and Sustainable Development/International Institute for Environment and Development (IIED)

Henríquez K (n.d.) Perspectivas de la Industria Minera del Oro en El Salvador. San Salvador, CEICOM

Hernández JR, Ávila JE, Hernández KE, Ochoa EA, Sandoval MA, Anzora MC (2011) Enfermedad renal en cuatro zonas de uso agrícola de El Salvador: prevalencia, factores de riesgo e indicadores asociados de disfunción hepática. San Salvador. Universidad Doctor Andres Bello

Ibarra ÁM, Jarquín UC, Rivera FJ (2005) Hacia la gestión sustentable del agua en El Salvador. Propuestas Básicas para Elaborar una Política Hídrica Nacional. San Salvador, Unidad Ecológica Salvadoreña (UNES)/Caritas

ICMM (2011) Mining: partnership for development toolkit. International Council on Mining and Metals, London

Kingsbury D (2014) Gold, water and the struggle for Basic Rights in El Salvador. Oxfam, Melbourne

Larios D (2008) La minería metálica y su impacto en El Salvador. San Salvador, CEICOM

Larios D, Cortes C, Cruz H (2011) Arsénico en aguas superficiales y subterráneas de las zonas mineras de los departmentos de La Unión y Morazán, El Salvador (Reporte Interno). San Salvador, CEICOM

MARN (2012) Informe técnico sobre muestreo de agua superficial realizado al rio San Sebastián, cantón San Sebastián, municipio de Santa Rosa de Lima (Mayo y julio de 2012). San Salvador. Gerencia de Cumplimiento Ambiental, Dirección General de Evaluación y Cumplimiento

MARN (2013) Determinación de sustancias tóxicas en agua y suelo, en areas expuestas a desechos de mineria metálica en el nororiente de el Salvador (5 Julio, 2013). San Salvador. Dirección General de Evaluación y Cumplimiento

Mckinley A (2012) Mitos y Realidades de la Mineria de Oro en Centroamerica. San Salvador, Caritas El Salvador

MINEC (2011) Evaluación Ambiental Estratégica (EAE) del sector minero metálico de El Salvador. Servicios de consultoría, Concurso publico internacional N° CPI—02/AECID/2010. San Salvador, Unidad de Cooperación Externa

MINEC (2012) Informe de inspección realizada en el área de la mina San Sebastián (1ro de Marzo, 2012). San Salvador. Ministerio de Economia

MINSAL (2015) Registered cases of illness in the Oriental región. San Salvador, Ministerio de Salud. Retrieved September 21, 2015 from the SIMMOW database

Monroy L (2008) Tóxicos en siete ríos de San Miguel y Morazán. La Prensa Gráfica. San Salvador. Retrieved March 30, 2016 from http://forodelagua.org.sv/noticia/2013/01/toxicos-en-siete-rios-de-san-miguel-y-morazan

Moreno R (2005) El marco jurídico para la privatización del agua en El Salvador. Brot für die Welt, Berlin

Nolasco S (2012) Impactos San Sebastian, El Salvador (Ficha para OCMAL). CEICOM, San Salvador. Retrieved July 17, 2015 from http://www.conflictosmineros.net/agregardocumento/publicaciones-ocmal/impactossansebastianelsalvador/detail

PDDH (2016) Informe Especial de la Procuraduria para la Defensa de los Derechos Humanos sobre el Legado de la Mina San Sebastian y sus Impactors en la Vida de las Poblaciones Afectadas. PDDH, San Salvador

Pepper M, Roche CP, Mudd GM (2014) Mining legacies—understanding life-of-mine across time and space. Paper presented at the Life-Of-Mine Conference, Brisbane (July)

Power T (2008) Metals mining and sustainable development in Central America: an assessment of benefits and costs. OXFAM AMERICA, Washington

Rosser A (2006) The political economy of the resource curse: a literature survey. IDS Working Paper 268. Brighton, IDS

SEC (2004) Commerce Group Corp, Form 10-K/A. Securities and Exchange Commission, Washington. Retrieved December 21, 2016, from https://www.sec.gov/Archives/edgar/data/109757/000114037705000136/0001140377-05-000136.txt

Telmer K, Veiga M (2008) World emissions of mercury from artisanal and smallscale gold mining and the knowledge gaps about them. Final draft, paper prepared for UNEP/ FT, Rome

UNEP (2008) The challenge of meeting mercury demand without mercury mining: an assessment requested by the Ad Hoc Open-Ended Working Group on Mercury. Concorde East/West SPRL, Geneva

UNEP (2009) Assessment report—excess mercury supply in Latin America and the Caribbean, 2010–2050. UNEP Chemicals, New York

USGS (2012) Mineral Commodity Summaries. January. Washington, USGS

Vallejo M (n.d.) La aventura subterránea de los "güiriseros" es grande. Diario de Oriente, La Union. Retrieved 30 May 2016, from http://www.elsalvador.com/DIARIOS/ORIENTE/2003/06/20/ACTUALIDAD/nota5.html

Van der Ploeg F (2011) Natural resources: curse or blessing? Journal of Economic Literature 49(2):366–420

Veiga M (1997) Mercury in artisanal gold mining activities in Latin America: facts, fantasies and solutions. Paper presented at UNIDO—Expert Group Meeting—Introducing new technologies for abatement of global mercury pollution deriving from artisanal gold mining, Vienna, Austria, July 1–3

Veiga MM, Baker RF (2004) Protocols for environmental and health assessment of mercury released by artisanal and small-scale gold miners. Global Mercury Project, UNIDO, Vienna

Velasco P (2000) The mineral industries of Central America: Belize, Costa Rica, El Salvador, Guatemala, Honduras, Nicaragua and Panama. Washington, USGS. Retrieved December 21, 2016 from https://minerals.usgs.gov/minerals/pubs/country/2000/camer00.pdf

WHO/UNICEF Joint Monitoring Program for Water Supply and Sanitation (2014) Estimates on the use of wáter sources and sanitation facilities in El Salvador. New York, WHO/UNICEF. Retrieved July 30, 2016 from http://www.wssinfo.org/

World Health Organization (n.d.) Minimum water quantity needed for domestic uses. Technical Note No. 9. WHO/SEARO, New Delhi

Chapter 5
Reform, Awareness, Prevention and Remediation Strategies

Keywords International best practice standards in mine closure · El Salvador mining law · State capacity · SODIS method · AMD remediation · Borax method · Participatory development · Post mining land use · Sustainable development

The findings in the previous chapter demonstrate that San Sebastian suffers from a number of legacies that, if untreated, will continue to adversely affect the district's environment and its inhabitants. Given that the owner of the mine is unwilling to take responsibility for remediating these legacies, it is the government of El Salvador's responsibility to do so. This will need concerted action at various levels if further damage is to be avoided.

5.1 Legislative Reforms Needed to Conform to International Best Practice in Mine Closure

To fulfill the requirements of international best practice it is important that legislators start with the premise that upon mine closure:

> Future public health and safety are not compromised, that the after-use of the site is beneficial and sustainable to the affected communities in the long term and that adverse socio-economic impacts are minimised and socio-economic benefits maximised (Whitbread-Abrutat et al. 2013: 638).

These expectations can only be met if the Salvadorean mining code is reviewed and strengthened. This includes the elimination of wording that allows companies to start working on closure activities *after* mining operations are completed. The current law clearly does not follow the principle that closure is a process that starts when mining commences and instead incentivizes companies to use emergency or patch up technologies. Planning for closure at the beginning of operations, on the other hand, incentivizes companies early in the operational stage to use procedures

© The Author(s) 2017
V. Pacheco Cueva, *An Assessment of Mine Legacies and How to Prevent Them*,
SpringerBriefs in Environmental Science, DOI 10.1007/978-3-319-53976-8_5

and technologies that are designed to leave little or no negative legacies at the end. Three examples of the lack of planning for closure are:

- There is no legal requirement for an environmental pre-mining baseline. Without such baseline it is difficult to determine what constitutes rehabilitation of the environment when closure begins.[1]
- The law does not define what constitutes successful mine closure and relinquishment procedures.[2]
- The requirements for closure and remediation are optional and not a pre-requisite for all mines.

In terms of financing closure the current law does not reflect remediation costs and requirements. The current law requires a bond of no less than US$50,000 but not exceeding US$300,000 per km^2 of mining concession to be used for environmental problems occurring at the end which effectively makes it a closure bond. From experience elsewhere it is clear that even the larger amount required by the law is not enough to remediate legacies of a large magnitude. Research conducted by the IFC (2002: 2) "indicates that medium-size open pit and underground mines operating in the past 10–15 years cost US$5–15 million to close, while closure of open pit mines operating for over 35 years, with large waste and tailings facilities, can cost upwards of $50 million". In 2004, the concession in San Sebastian covered 2.1 km^2 so the bond should have been a figure between US$100,000 and 600,000 at the rate stipulated by the law. The actual figure paid by company was US$14,428[3] which is far less than the minimum figure stipulated by the legislation.[4] It is also counter-intuitive that the law allows companies to choose paying the bond instead of spending on measures that prevent on-going environmental damage.[5] This effectively incentivizes companies to switch to mechanisms that have very little effectiveness in terms of environmental protection. The ideal then is that there should be no substitute for paying for ongoing environmental rehabilitation. An environmental bond, in this case, should only be used in emergency scenarios and not be solely based on closure costs estimates from the proyect's EIS but must be re-evaluated over the mine's lifecycle to take into account costs not anticipated in the EIS.[6]

---

[1]See Article 24 of the Mining Law.

[2]A closure plan (*Plan de Abandono*) is defined in the law thus: A plan is a document, duly approved by the Ministry of Environment and Natural Resources, containing actions and deadlines that commits the holder of an exploration concession or exploitation of minerals or hydrocarbons to restore the environment or implement compensatory measures, if any, after completing exploration work or exploitation (Legislative Decree No. 566, 4th October 2001, published in the National Gazette, Number 198, Volume 353, 19th October 2001).

[3]Ministry of Environment and Natural Resources, Resolution Number 493, 2002.

[4]See Article 14 of the Mining Law.

[5]See Article 14 of the Mining Law and Article 83 of the Environmental Law.

[6]See Article 29 of the Mining Law.

5.2 Prevention and Awareness

Despite the fact that the El Salvador's environmental law contains many channels to prevent environmental damage, it is clear from the findings in San Sebastian that the ability of government agencies in charge of monitoring and enforcing the legislation remains limited. It must be said that the San Sebastian mine was not a large operation when compared to some of the current mines operating elsewhere in Latin America. If, at any time in the future, the mining moratorium is lifted and medium to large scale mining companies seek licenses to operate in the country, the government is well advised to greatly improve its own regulatory capacity to be able to cope with the increased complexity of larger mining operations. This also includes increasing the government's capacity to analyse the financial operations of mining companies so that the appropriate taxes and royalties are paid. Apart from finding that the government has a lack of regulatory capacity this study also found that there is no clear and transparent mechanism by which the mining royalties allocated to the municipalities travel from the company to the central government and onward to its destination. In keeping with the principles of upholding transparency at all levels and implementing sustainable development, it would be advisable if the municipal government made that amount public upon receipt of funds from the central government and to make these funds available only to assist in the development efforts of the people most affected by mining.

Prevention also means taking action to redress mining legacies. A shift from the government's current policy position of upholding a moratorium on mining to a position of actively closing down the country's 15 abandoned and orphaned mines will help in this regard.[7] The mining moratorium in El Salvador means that nothing related to mining is prioritized by the government, including remediation of legacy sites. This administrative paralysis creates uncertainty in both the private and civil society sectors and decreases the capacity of the government to learn about mine closure requirements. This shift in policy position is equally important for the country's anti-mining movement that calls for a complete ban on mining.[8] Closing down the country's abandoned and orphaned mines in order to prevent further damage like the one occurring in San Sebastian is a task that needs to be carried out regardless of the presence or absence of an active mining

---

[7]The government has conducted some work to catalogue all of these sites. See Domínguez et al. (2015).

[8]The environmental movement's awareness strategy highlights the many harmful risks that mining entails. The government's adoption of the mining ban and the popular disproval of the mining industry mean that this awareness strategy is very successful but until 2016 little had been done to add mine closure considerations to the strategy. See for instance http://www.stopesmining.org/j25/.

industry.[9] In this regard it is important to add that since the principle of polluter pays informs environmental law in El Salvador and the government is unable to cover all the costs of closing all abandoned mines there is a strong argument for the government to pursue litigation against companies such as CGC for environmental damages caused by them.[10] In this instance litigation sends a clear signal to industry that the government is serious about its environmental and human rights commitments thus preventing rouge companies from not adhering to the law or from investing in the country. El Salvador's sole environmental court is underutilized with not a single case of environmental damage involving mining companies under consideration by its officials so far.[11]

One last action in terms of prevention is the establishment of a mining legacies fund similar to the one operating in Canada under the auspices of the Ministry for the Environment.[12] The aim of the proposed fund is to prevent damage from legacies that closure procedures could not prevent and should be kept separate from the already existing environmental bond required of companies for emergency environmental damages. Whereas the latter deals with short term problems that may arise during the mine lifecycle, the former handles those problems that occur long after the closure of mining and that may happen despite the implementation of the best mine closure procedures. Money for this fund should initially be raised by the government of El Salvador in conjunction with international financial institutions to help remediate mines like the one in San Sebastian. If the mining moratorium is ever lifted then all those companies with a mining license should contribute to the fund.

---

[9]The prevention of further damage is obvious when the presence of lead in San Sebastian soils is considered. Since MARN (2013) made clear that the levels of lead in the soil in some areas of San Sebastian are higher than the established safe limits, the government's next step is to test people living in those areas to determine the level of lead in their blood. The second step is to educate the population about how to deal with its presence, how to avoid it and how to recognize the symptoms associated with lead poisoning. The final step is to assist those living in those areas in cleaning up their environment and maintaining it that way. The same approach can be taken with all other heavy metals that are above the safe limit.

[10]The legal basis for the government to take such action is found in Environmental Law (1998), Articles 85 and 86 subsections c, d. g and h.

[11]Personal communication with the country's sole environmental judge, Mr. Samuel Lizama Morales (January 28th, 2016). Mining companies, on the other hand, have taken the government to court on at least two occasions in the past few years. The CGC case described earlier in this book is one. The other involves the Canadian/Australian company Pacific Rim. See more on this case here: https://www.theguardian.com/global-development/2016/oct/14/el-salvador-world-bank-tribunal-dismisses-oceanagold-mining-firm-250m-claim.

[12]http://www.ec.gc.ca/edf-fde/.

5.3 Remediating Legacies in San Sebatian

Out of all the main legacies identified during the course of the current study only those categorized as lost opportunities do not have possible avenues for remediation. The others will be treated separately in this chapter and will start with those relating to remediating the impacts of AMD on San Sebastian's surface and ground water, followed by improving access to non-polluted water and finding alternatives to artisanal and small scale mining. The chapter will not include a discussion on legacies such as land tenure uncertainty and health problems because, despite the best efforts, there is still not enough information to be able to draw a strategy to remedy them.

5.3.1 AMD Remediation

A discussion of remediation strategies for San Sebastian requires not only an examination of available techniques and methods but also a calculation of their costs so that the full economic impact of this type of legacy can be gained.

The US Environmental Protection Agency (EPA 2007) has produced a guide for those interested in calculating the global costs of AMD prevention and remediation. According to the Agency, remediation costs vary greatly depending on a number of factors including the degree of contamination, the size of the operation, the climate, the level of rehabilitation required and the soil types found at the site. In addition the science that analyzes the prevention and remediation of acid mine drainage is very broad, and the methods that have been developed at the international level offer a wide range of options to deal with this problem. Johnson and Hallberg (2005) provide a summary of the most recent techniques that he divides them into prevention and remediation techniques.[13]

Upon examining the site, it is obvious that a successful remediation strategy will need to include a mixture of prevention and remediation techniques. The objective of preventive techniques is to reduce the amount of water that reaches the waste and exposed rock in order to reduce chemical reactions in it. Taking into account the EPA costs guide and the available site details, it is thought that the most appropriate

---

[13]These techniques are subdivided into passive and active. The former include flooding/sealing of underground mines, underwater storage of mine tailings, land based storage in sealed waste heaps, the blending of mineral wastes, total solidification of tailings, the application of anionic surfactants and microencapsulation (coating). Remediation techniques include abiotic and bioremediation systems. Abiotic systems rely on minerals to neutralize acid discharge and evaporation. There are active systems that use aeration and lime and passive systems that can include the use of anoxic limestone drains. Bioremediation systems use aerobic bacteria to neutralize acid discharge. These are divided into passive and active systems. The former include off-line sulfidogenic bioreactors and the latter include aerobic wetlands, compost reactors, permeable reactive barriers, and packed bed iron-oxidation bio-reactors.

prevention strategy is the sealing of waste heaps and underground mines and that these could cost approximately US$70,000.[14]

However, techniques to prevent water flowing into the rock are not 100% effective, especially during the rainy season and on the sloping ground near the microcatchment of two local streams that flow into the San Sebastian River (Domínguez et al. 2015: 169). The expectation is that, even with prevention strategies in place, a certain amount of water will sometimes contact the mineralized rock, starting the process of AMD again, albeit with a smaller rate of flow. This is where remediation techniques can play a part. Of the techniques identified by Johnson and Hallberg (2005) bio-remediation is the most appropriate for this mine for reasons of costs and water flow. Bio-remediation can be active or passive. Passive systems generally use the activity of microorganisms and require higher initial costs, but maintenance costs over the long-term are lower, with an average of 10% of the initial construction costs. However, costs depend to a large extent on the volume of water required for treatment, the amount of material added to the water for treatment, the average annual amount of rainfall and the retention characteristics of the mine and surrounding rock (EPA 2007: 14).

Passive bio-remediation encompasses several methods but those consisting of wetlands can be used in places that exhibit low water flows. Judging from the close variation in rainfall in San Sebastian, it can be safely assumed that water treatment costs will be slightly higher in the rainy season than in the dry season, but the margin will not be very high. The water flow at the point of discharge, according to CEICOM (2010: 42) and Nolasco (2012: 4) is 180 l/min (47 gal/min) during the rainy season and 120 l/min (31.7 gal/min) in the dry season, which in general can be considered a low flow discharge. Therefore a remediation system based on wetlands may be the most appropriate for this site. The construction costs associated with such treatment can start at US$179,000[15] and the annual operating costs range from US$15,000 to US$52,000.[16]

One final but essential consideration is that the active participation of the local population is something that the above calculations do not take into account. Any prevention and remediation project in San Sebastian needs to include a local social component that must receive sufficient resources to improve the chances of success. In addition to assisting in the planning, construction and maintenance phases of

---

[14]The figure includes the use of diversion ditches and berms, collection and treatment with cover, composite soil cover and synthetic liner depending on the requirements of the site. But it does not include the costs of environmental permits, legal, training and other consulting costs, prevention and remediation of contamination already established in the environment, especially that caused by mercury and other heavy metals in soil, groundwater and living creatures. Therefore the rehabilitating of all the damaged ecosystems in San Sebastian will require a much larger investment than the one presented in this study.

[15]Based on a flow of 150 l/min (39.6 gal/min), which is the average between the dry and rainy season flows, as baseline and EPA's average capital costs that range from US$2900 to US$18,500 gal/min.

[16]Based on the EPA average annual operating costs range from US$120 to US$420 gal/min. The water flows are based on CEICOM's estimates.

such a project, the local inhabitants can provide valuable information on environmental conditions and assist in environmental monitoring, evaluation and conservation. Therefore it is important that remediation and other processes associated with closure of the mine follows the principles of participatory and inclusive development (Pacheco Cueva 2012).

5.3.2 Improving Access to Non-polluted Water

Although the situation of water scarcity is not particular to San Sebastian, this district represents a critical case because the population has little access to water sources such as rivers, streams and wells that traditionally satisfy water needs in rural areas. To remedy the situation Caritas El Salvador has been working to improve the access to non-polluted water to residents of San Sebastian by helping them build water tanks and training them in the use of techniques such as the SODIS method of water purification.[17] The same organization has raised awareness of the problems associated with not drinking enough water and using polluted water for agricultural purposes and for raising domestic animals and plants. However, an interview with a Caritas representative who has worked in these projects revealed that the organization cannot keep up with the demand for non-polluted water in San Sebastian and instead have focused their efforts on building the community's capacity to negotiate with the government's water authorities directly.[18] What has resulted from these efforts is that community leaders have come to realize that the supply of water through privatized means is not a solution for San Sebastian. The community is already paying exorbitant prices for its water through what is basically a system of private providers. The most economical solution for the inhabitants of San Sebastian is to convince government authorities to start planning a system that will provide potable water and sanitation for the whole district.[19] Of course, this system will not solve the problem of lack of non-polluted water for commercial agriculture but it will go a long in alleviating the costs of obtaining non-polluted water for human consumption in the district.

5.3.3 Finding Alternatives to Artisanal and Small Scale Mining

This complex legacy cannot be remediated. However a program of transition from mine dependency to a more sustainable form of livelihood, which may or may not

---

[17]http://www.sodis.ch/methode/index_EN.

[18]The Caritas PRA mentioned earlier in this book is part of those efforts.

[19]This has already been mentioned in the MARN San Sebastian study (2013). As of the end of 2016, a group of community leaders submitted a proposal for such a system and are awaiting a government response.

include mining, must be found for the people of San Sebastian. Domínguez et al. have suggested the introduction of sustainable livestock production and rock cutting and add that the former is only "viable for *guiriseros* who own land and have prior experience producing livestock in the district" (2015: 174–177). It should also be mentioned that this option faces additional problems such as the lack of quality soils and water that complies with national environmental standards.[20] Rock cutting takes advantage of the *güiriseros's* prior skills but will require some re-training, safety equipment and market outlets. In the same document Domínguez et al. discard the option of tourism development even though some of the *güiriseros* commented during the RRA interviews that the area has a high tourist attraction because it is close to Santa Rosa de Lima, a national tourist attraction in its own right. The *güiriseros* ideas may have some prospects. There are heritage mining projects in Chile, Bolivia and beyond where old mines have been transformed into tourist attractions and ex-miners now act as guides explaining how mining was conducted in the past to paying visitors. While it is true that the Sewell mine in Chile and the mining town of Potosí in Bolivia have achieved UNESCO world heritage status and currently attract international tourists, a project to convert the San Sebastian mine into a tourist site would have to begin as a modest national attraction. Also, this initiative obviously requires that the environment of the area be in a process of recovery.

Regardless of the initiative that best fits San Sebastian, the transition to other economic alternatives requires the incorporation of the *güiriseros* in the decision making process. A strategy to engage the *güiriseros* should therefore include the following elements:

- It is necessary that the *güiriseros*' activities be regulated and the use of mercury halted in favour of alternatives such as the Borax method.[21] In addition to this, the creation of a work culture that emphasizes occupational safety should be encouraged.
- Self-organizing the *güiriseros* and their families must be supported so that they can be involved in joint decision-making and be the main actors in determining their own development as a community.
- The creation of a larger organization involving representatives of the San Sebastian community, the environmental movement, local businesses and the government to generate proposals for environmental recovery and alternative socio-economic development for the district must be supported.

A final remark is that any alternatives chosen have more chances of success if these are be able to generate at least the US$0.72 million that ASGM activity currently generates.

---

[20]To this effect, an official of the Ministry of Agriculture stated during one of the RRA interviews that "people in the area do not handle much cattle because there is no access to water in the area and it is well known that the water in the area is contaminated".

[21]See: https://www.911metallurgist.com/blog/mercury-free-gravity-borax-method-gbm.

References

CEICOM (2010) Análisis de la calidad de agua y su relación con la salud y calidad de vida de los pobladores del Río San Sebastián, en la zona de minas San Sebastián. San Salvador, CEICOM

Domínguez JP, Moya M, Rodríguez E, Panameño P, Linares J (2015) Evaluación final de riesgos y propuesta de medidas de remediación en 15 pasivos ambientales mineros de El Salvador. San Salvador, Fundación Maquilishuatl, FIAES, MINEC

EPA (2007) Costs of remediation at mine sites. Environmental Protection Agency, Washington

IFC (2002) It's not over when it's over: mine closure around the world. International Finance Corporation/World Bank, Washington

Johnson DB, Hallberg KB (2005) Acid mine drainage remediation options: a review. Sci Total Environ 338:3–14

MARN (2013) Determinación de sustancias tóxicas en agua y suelo, en areas expuestas a desechos de mineria metálica en el nororiente de el Salvador (5 Julio, 2013). San Salvador. Dirección General de Evaluación y Cumplimiento

Nolasco S (2012) Impactos San Sebastian, El Salvador (Ficha para OCMAL). CEICOM, San Salvador. Retrieved July 17, 2015 from http://www.conflictosmineros.net/agregardocumento/publicaciones-ocmal/impactossansebastianelsalvador/detail

Pacheco Cueva V (2012) Foundations, trusts and funds in near mine closure and post-closure environments: a case study from Bolivia. In: Tibbett M, Fourie AB, Digby C (eds) Proceedings seventh international conference on mine closure. Australian Centre for Geomechanics, Perth, pp 747–758

Whitbread-Abrutat P, Kendle A, Coppin N (2013) Lessons for the mining industry from non-mining landscape restoration experiences. In: Tibbett M, Fourie AB, Digby C (eds) Proceedings eight international conference on mine closure. Australian Centre for Geomechanics, Perth, pp 625–640

Conclusion

This study analysed and quantified, as far as the available information allowed, the socioeconomic legacies of the abandoned San Sebastián mine in eastern El Salvador. Thanks to the review of numerous studies carried out in relation to AMD contamination of the local river it was possible to establish that the damage is severe and its impacts are felt well beyond the mining district. It is estimated that the initial costs of prevention and remediation can be US$70,000 and US$180,000 respectively and that maintenance costs can range from US$15,000 to US$52,000 per annum.

The impact of access to non-polluted water for human consumption is also a serious legacy. Rural communities in El Salvador suffer from severe water short-ages, but in the case of San Sebastián, the inhabitants cannot make use of the natural sources of water. To fill the gap between the water they need to survive and the available water, the residents of San Sebastian have to devote up to 20% of their income in obtaining water. This, in turn, inhibits their chances of enjoying a more prosperous life.

Apart from environmental legacies, the mine also left a number of socio-economic legacies. These include the lost opportunities due to the cessation of mine operations worth over US$200,000 per annum for the district of San Sebastian alone. In relation to land tenure, there is great deal of uncertainty for the families living on what was CGC property and this, in turn, inhibits their economic opportunities.

Since CGC abandoned the mine, ASGM activities have grown to compensate for the lost economic opportunities described above. These activities impact the environment and may be having long term effects on the health of the *güiriseros* and their families. In addition, every year many *güiriseros* die or are injured in accidents as a result of the poor working conditions found inside the underground tunnels and the lack of appropriate safety measures and equipment.

Some of these legacies could have been prevented or ameliorated if a strong legal framework to deal with mining problems were present to protect both inhabitants and their environment. The study found that in many important respects the current legal framework does not meet international best practice when it comes to mine closure and relinquishment requirements. Also, the government's

© The Author(s) 2017
V. Pacheco Cueva, *An Assessment of Mine Legacies and How to Prevent Them*,
SpringerBriefs in Environmental Science, DOI 10.1007/978-3-319-53976-8

transparency mechanisms to track the proceeds of mining are completely absent and the ones that are present are difficult to access. Much reform or a legislated ban on mining is therefore needed.

The study also found that state capacity to plan closure, prevent mining risks, implement legislation and ensure compliance with the law is very weak. One instance of sheer lack of capacity relates to health assessments. It is not uncommon for health authorities not to have the necessary information and studies to ascertain what may be affecting the population even when harmful substances such as heavy metals derived from mining operations are present in the environment.

The study provided strategies for awareness, prevention and remediation. The awareness strategies provided aimed to assist the inhabitants of San Sebastian, government authorities and civil society in El Salvador and beyond realise the risks involved in extractive projects and either terminate or treat them. The aim of the prevention strategies was to avoid further damage from occurring while that of rehabilitation strategies was to show that some rehabilitation of San Sebastian is possible. However, the study also aimed to show that rehabilitation, when possible, requires not only technical and financial resources but also political will, local participation and a long term view of development.

This brings us to the other objective of the study which was to contribute to the growing body of knowledge about mining legacies. In this regard, the study makes three contributions. The first one has to do with showing that legacies involve very complex social, economic and environmental phenomena and their remediation, if possible, requires an equally complex set of solutions. The second contribution highlights how environmental negligence by a private operator becomes, upon abandonment, a cost that has to be borne by society. These two contributions were given priority to demonstrate that non-renewable extraction projects, under certain circumstances, may be so costly for society that they should not be considered as an economic development option.

The last contribution emphasises that, if non-renewable extraction project are under consideration as an economic development option, then communities and governments should make their closure the most important long term planning consideration. Long term planning is significant because most rural economies are already highly dependent on a single or few commodities. Closure planning, therefore needs to consider the factors required for a successful economic transition from being dependent on a single or narrow band of resources to becoming more economically diverse while improving the sustainability of current, non-extractive economic activities. This type of planning also needs to consider appropriate monitoring mechanisms that tell communities if and when social and economic goals are achieved. It is therefore essential that project closure planning and broader economic planning is integrated and supports existing drivers of the formal economy. As a final point, mining closure plans should be subject to existing global standards like the ones mentioned at the beginning of this book, especially as they apply to long term planning, transparency, human rights issues, security, environmental management, local participation and sustainability.

Appendix A

IFC Guidelines for Mine Closure and Post Closure (2007).

Planning. Closure and post-closure activities should be considered as early in the planning and design stages as possible. Mine sponsors should prepare a Mine Reclamation and Closure Plan (MRCP) in draft form prior to the start of production, clearly identifying allocated and sustainable funding sources to implement the plan. For short life mines, a fully detailed Mine Reclamation and Closure Plan (with guaranteed funding) as described below should be prepared prior to the start of operations. A mine closure plan that incorporates both physical rehabilitation and socio-economic considerations should be an integral part of the project life cycle. The MRCP should address beneficial future land use (this should be determined using a multi- stakeholder process that includes regulatory agencies, local communities, traditional land users, adjacent leaseholders, civil society and other impacted parties), be previously approved by the relevant national authorities, and be the result of consultation and dialogue with local communities and their government representatives. The closure plan should be regularly updated and refined to reflect changes in mine development and operational planning, as well as the environmental and social conditions and circumstances. Records of the mine works should also be maintained as part of the post-closure plan.

Monitoring. Closure and post closure plans should include appropriate aftercare and continued monitoring of the site, pollutant emissions, and related potential impacts. The duration of post-closure monitoring should be defined on a risk basis; however, site conditions typically require a minimum period of five years after closure or longer.

Timing. The timing for finalization of the MRCP is site specific and depends on many factors, such as potential mine life, however all sites need to engage in some form of progressive restoration during operations. While plans may be modified, as necessary, during the construction and operational phases, plans should include contingencies for temporary suspension of activities and permanent early closure and meet the [established] objectives for financial feasibility and physical/chemical/ecological integrity.

Financial Feasibility. The costs associated with mine closure and post-closure activities, including post-closure care, should be included in business feasibility

© The Author(s) 2017
V. Pacheco Cueva, *An Assessment of Mine Legacies and How to Prevent Them*, SpringerBriefs in Environmental Science, DOI 10.1007/978-3-319-53976-8

analyses during the planning and design stages. Minimum considerations should include the availability of all necessary funds, by appropriate financial instruments, to cover the cost of closure at any stage in the mine life, including provision for early, or temporary closure. Funding should be by either a cash accrual system or a financial guarantee. The two acceptable cash accrual systems are fully funded escrow accounts (including government managed arrangements) or sinking funds. An acceptable form of financial guarantee must be provided by a reputable financial institution. Mine closure requirements should be reviewed on an annual basis and the closure funding arrangements adjusted to reflect any changes.

Appendix B

Mine closure bond requirements according to World Bank guidelines (Da Rosa 1999).

Closure costs. Financial assurances must cover the operator's cost of reclamation and closure as well as redress any impacts that a mining operation causes to wildlife, soil, and water quality. The bond should also cover the cost of a post-closure monitoring period. To accurately compute the level of financial assurance, reclamation and mitigation activities should be clearly spelled out in the operation plan. In addition, the bond should cover the costs of addressing impacts that stem from the operator's failure to complete reclamation, such as the need for long-term treatment of surface and groundwater, environmental monitoring and site maintenance. During mining, assurance levels should be subject to periodic reviews, in order to allow regulators to adjust operators' assurance amounts upward or downward as clean-up needs, environmental risks, or economic factors dictate.

Liquidity. All forms of financial assurance should be reasonably liquid. Cash is the most liquid asset, but high-grade securities, surety bonds and irrevocable letters of credit can serve as acceptable forms of assurance. However, assets that are less liquid, particularly the mine operator's own property or equipment should not be considered adequate assurance, since these items may quickly become valueless in the event of an operator default or bankruptcy.

Accessibility. Financial assurances should be readily accessible, dedicated and only released with the specific assent of the regulatory authority, so that regulators can promptly obtain funding to initiate reclamation and remediation in case of operator default. Forms of financial assurance should be payable to regulators, under their control or in trust for their benefit, and earmarked for reclamation and closure. Further, such financial assurances must be discreet legal instruments or sums of money releasable only with the regulatory authority's specific consent.

Timeliness. For their part, regulators must obtain financial assurance up front before a mine project is approved. While regulators, as determined by their periodic reviews, must have the authority to secure financial assurance during the course of mining, waiting until late in the mining process to obtain substantial assurance is unwise, since reduced cash flows at this stage may make it difficult for operators to secure bonding from a surety, bank, or other guarantor.

© The Author(s) 2017
V. Pacheco Cueva, *An Assessment of Mine Legacies and How to Prevent Them*,
SpringerBriefs in Environmental Science, DOI 10.1007/978-3-319-53976-8

Healthy guarantors. To assure that guarantors have the financial capacity to assume an operator's risk of not performing its reclamation obligations, regulators must carefully screen guarantors' financial health before accepting any form of assurance. Any risk sharing pools should also be operated on an actuarially sound basis. Regulators should require periodic certification of these criteria by independent, third parties.

Public involvement. Since the public runs the risk of bearing the environmental costs not covered by an inadequate or prematurely released bond, the public must be accorded an essential role in advising authorities on setting and releasing of bonds. Therefore, regulators must give the public notice and an opportunity to comment both before the setting of a bond amount and before any decision on whether to release a bond.

No substitute. Any financial assurance should not be regarded as a surrogate for a company's legal liability for clean-up, or for the regulators' applying the strictest scrutiny and standards to proposed mining plans and operations. Rather, a financial assurance is only intended to provide the public with a buffer against having to shoulders costs for which the operator is liable.

Bibliography

Baker Matta M, McKinnie D, Barraza E, Sericano J (2002) Hurricane Mitch Reconstruction/Gulf of Fonseca Contaminant Survey and Assessment. National Oceanic and Atmospheric Administration, Maryland

Chambers R (1980) Rural development: putting the last first. Harlow, England

Clark A, Clark J (2005) An international overview of legal frameworks for mine closure. International Development Research Center (IDRC), Ottawa

Cochilco/Chilean Copper Commision and UNEP (2001) Abandoned mines—problems, issues and policy challenges for decision makers. Santiago de Chile and Paris, Cochilco and UNEP

Cochilco/Chilean Copper Commission (2002) Research on Mine Closure Policy. International Institute for Environment and Development/World Business Council for Sustainable Development, Mining, Minerals and Sustainable Development, London, January, No. 44. Retrieved June 20, 2016 from http://pubs.iied.org/pdfs/G00541.pdf

Da Rosa C (1999) Financial planning for mine closure. Mining Environ Manag 7(2):10–13

IFC (2006) Policy on social and environmental sustainability. International Finance Corporation/World Bank, Washington

Karunananthan M, Spronk S (2015) Water at the heart of El Salvador's struggle against neoliberalism. Ottawa, Blue Planet Project

Larios D, Guzman H, Mira E (2008) Riesgos y posibles impactos de la mineria metalica en El Salvador, vol 63. Estudios CentroAmericanos, San Salvador, Numero 711–712

Macdonald C, McGuire G, Weston H (2006) Integrated closure planning review: literature review. ICMM, London

MARN (2010) Informe de Calidad de Agua de los ríos de El Salvador. Ministerio de Ambiente y Recursos Naturales, San Salvador

USGS (n.d.) How does mine drainage occur? Retrieved November 15, 2015 from http://www.usgs.gov/faq/categories/9816/2573

Veiga M, Roberts S, Peiter C, Sirotheau G, Barreto ML, Ezequiel G (2000) Filling the void: the changing face of mine reclamation in the Americas. Department of Mining and Mineral Process Engineering, University of British Columbia and CETEM (Centro de Tecnologia Mineral), Vancouver and Rio de Janeiro

World Bank (2008) Guidance notes for the implementation of financial surety for mine closure. World Bank, Washington

World Bank (2010) Towards sustainable decommissioning and closure of oil fields and mines: a toolkit to assist government agencies (Version 3.0). World Bank, Washington

© The Author(s) 2017
V. Pacheco Cueva, *An Assessment of Mine Legacies and How to Prevent Them*, SpringerBriefs in Environmental Science, DOI 10.1007/978-3-319-53976-8